THE PHYSICS OF ELEMENTARY PARTICLES

INVESTIGATIONS IN PHYSICS

Edited by EUGENE P. WIGNER and ROBERT HOFSTADTER

1. Ferroelectricity by E. T. JAYNES
2. Mathematical Foundations of Quantum Mechanics by JOHN VON NEUMANN
3. Shell Theory of the Nucleus by EUGENE FEENBERG
4. Angular Momentum in Quantum Mechanics by A. R. EDMONDS
5. The Neutrino by JAMES S. ALLEN
6. Photonuclear Reactions by L. KATZ, R. SKINNER, and A. S. PENFOLD (to be published)
7. Radiation Damage by D. BILLINGTON and J. CRAWFORD (to be published)
8. Nuclear Structure by LEONARD EISENBUD and EUGENE P. WIGNER
9. The Physics of Elementary Particles by J. D. JACKSON

THE PHYSICS OF ELEMENTARY PARTICLES

By J. D. JACKSON

PRINCETON, NEW JERSEY
PRINCETON UNIVERSITY PRESS
1958

London: Oxford University Press
L.C. Card 58-13935

Printed in the United States of America

PREFACE

My aim in writing this little book has been to present an introductory account of elementary particle physics with a minimum of formal apparatus. The discussion will therefore be somewhat superficial in places, but I hope that a lack of depth here and there will be made up for in an ease of reading that a more formal treatment would prevent. My purpose is not to educate the educated, but to provide an introduction to the subject from which the reader can go to the literature for a deeper probing at the details.

The material is divided into three roughly equal parts. The first part is on low-energy pion physics, and covers the interaction of pions with nucleons at energies below 300 or 400 Mev. The second part deals with the phenomena of K-mesons and hyperons and their interactions with nucleons. The final part treats decay processes, especially the beta decay interactions, and summarizes the present status in that rapidly developing field. Only the simplest fundamental processes have been treated. Consequently many active fields of particle physics have been slighted. Antinucleon phenomena and nucleon-nucleon forces are two that come immediately to mind. My excuse, if I should need one, is that these and other phenomena are sufficiently complicated that they do not at the present time add to our fundamental understanding of particles and their interactions. In an elementary account it is best to stick to the simplest things.

This book had its origin in a series of lectures given at the Summer Seminar of the Theoretical Physics Division of the Canadian Association of Physicists, in Edmonton, Alberta, during August 1957. It is a pleasure to acknowledge the stimulus of that occasion in the preparation of the present work.

In a field of physics that is changing every instant it is senseless to hope that a book written today will endure much past tomorrow, but I hope that this volume will serve for a short time as a useful introduction to a fascinating subject.

J. D. Jackson

December 1957

A few changes and additions were made in galley proof to incorporate some recent results. References added at that time are denoted by a dagger or other symbol rather than by number.

J. D. J.

May 1958

TABLE OF CONTENTS

PART I

THE INTERACTION OF PIONS AND NUCLEONS

PART II

K-MESONS AND HYPERONS

PART III

DECAY PROCESSES

LIST OF FIGURES

PART I

THE INTERACTION OF PIONS AND NUCLEONS

CHAPTER 1

Introductory Discussion of Pions and Nucleons

Over twenty years ago Yukawa proposed the meson as the quantum of the nuclear force field, in analogy with the photon as the quantum of the electromagnetic field. In subsequent years physicists have come to a reasonably good understanding of the behavior of pions and nucleons, at least at low energies. The path was not particularly straight. For example, there was a detour involving the muon, discovered long before the pion, and having nothing to do with the nuclear force field. And there was great preoccupation with the specific problem of nuclear forces, still in a somewhat unsatisfactory state. One can make a poor analogy with electromagnetism, where the original research dealt with the law of force, Coulomb's law, but the complete nature of the electromagnetic field was not known until Thomson and Compton scattering had been studied in detail. In pion physics the nuclear force problem is complicated by the great strength of the interaction and by the existence of strongly interacting heavy mesons. On the other hand, the pion-nucleon interaction (scattering and photoproduction) is fortuitously free from too severe complications until the energy is high enough that inelastic processes and heavy mesons enter importantly. This is chiefly because the so-called "3,3 resonance" dominates the interaction at low energies. The fortunate "cleanness" of the pion-nucleon problem means that we learn a great deal from the elementary acts of scattering and photoproduction from hydrogen (and deuterium). It is these processes that will be treated in Part I. The nuclear force problem will be mentioned only briefly.

Our discussion will follow the lines of the "static" model, treated in detail by Chew[1] in his excellent review article based largely on the work of Chew and Low.[2,3]

1.1. Properties of Pions

Pions are unstable particles of mass around 270 times the electron mass. There are three different pions: positively charged and negatively charged, both with charge equal in magnitude to the electronic charge, and neutral. The charged pions decay into a muon and a neutrino, while

[1] G. F. Chew, "Theory of Pion Scattering and Photoproduction", *Encyclopedia of Physics, Vol.* 43, Springer-Verlag, Berlin (to be published).
[2] G. F. Chew and F. E. Low, *Phys. Rev. 101*, 1570 (1956).
[3] *Ibid.*, 1579 (1956).

the neutral pion decays into two light quanta. The chief properties of pions and muons are displayed in Table 1. The spin of the (charged) pion is inferred from detailed balancing arguments on the reaction $p + p \rightarrow d + \pi^+$, while the decay of the neutral pion into two gamma rays proves that its spin is even (higher values than zero are unlikely). The intrinsic parity of the charged pions can be inferred from the large probability[4] for the

Table 1. Properties of Pions and Muons

	$\pi^\pm$	π^0	$\mu^\pm$
Charge	$\pm e$	0	$\pm e$
Mass	$273.3 \pm 0.1\ m_e$	$264.3 \pm 0.3\ m_e$	$206.9 \pm 0.1\ m_e$
Decay Process	$\pi^\pm \rightarrow \mu^\pm + \nu$	$\pi^0 \rightarrow \gamma + \gamma$	$\mu^\pm \rightarrow e^\pm + \nu + \nu$
Mean Lifetime	$2.56 \pm 0.05 \times 10^{-8}$ sec	$\tau < 4 \times 10^{-16}$ sec	$2.22 \pm 0.02 \times 10^{-6}$ sec
Spin	0	0	$\frac{1}{2}$
Intrinsic Parity	Odd	Odd	—

capture process, $\pi^- + d \rightarrow 2n$. Throughout the book units such that $\hbar = c = 1$ will be employed, and in addition, in Part I the energy of the charged pion (139.6 Mev) will serve as the energy unit, its Compton wavelength (1.41×10^{-13} cm) as the unit of length, etc. We will ignore the mass differences between charged and neutral pions and between neutron and proton. In these units the total energy of a pion of momentum q is $\omega_q = \sqrt{q^2 + 1}$. The rest mass of a nucleon is $M = 6.72$.

1.2. Form of the Pion-Nucleon Interaction

Without considering the specific properties of pions, certain qualitative implications about the pion-nucleon interaction can be drawn from the known nuclear two-body system, assuming that the pions are the major contributors to the nuclear force. First of all, we know that the nuclear two-body force is spin dependent—the bound n–p system is in a triplet spin state; the singlet spin state is unbound. Secondly, the deuteron has a quadrupole moment, implying the existence of a non-central force, the "tensor" force. And lastly, the two nucleon forces appear to be "charge

[4] See Sec. 28 of H. A. Bethe and F. de Hoffmann, *Mesons and Fields, Vol. II*, Row, Peterson, Evanston, Ill., 1955, for the detailed arguments on spins and parities.

independent", e.g. the singlet n–p, p–p, and n–n forces are equal, apart from Coulomb effects, at least at low energies.

If we consider the forces as due to the basic process of emission and absorption of single pions, we must try to write down an interaction between pions and nucleons that is linear in the meson field ϕ. Since the two-body force is spin dependent, the interaction must involve the nucleon's spin. Forgetting about charge for the moment and describing the nucleon by a classical spherically symmetric source density $\rho(\vec{r})$, with a spin angular momentum $\vec{\sigma}$, one is essentially forced into writing down an interaction Hamiltonian density of the form:

$$H_{\text{int}} \sim \rho(\vec{r})\, \vec{\sigma} \cdot \vec{\nabla}\phi. \tag{1.1}$$

More complicated forms are possible, of course, but this is the simplest one coupling the spin to the pion field. It is easy to show[5] that such an interaction leads to a two-body force involving $\vec{\sigma}_1 \cdot \vec{\sigma}_2$ and S_{12}, the tensor force operator. Thus the potentiality of satisfying the first two requirements of the nuclear force problem is present in an interaction of the form shown in (1.1). Another way of obtaining the Hamiltonian density (1.1) is to observe that a linear coupling of the pseudoscalar pion field ϕ (odd intrinsic parity) to the nucleon must involve a nucleon variable which is a pseudotensor of some rank so that the interaction can be a true scalar. Nonrelativistically, the only appropriate variable is the nucleon's spin, an axial vector. The gradient of the pion field forms another axial vector, and its scalar product with the nuclear spin yields a scalar under rotations and inversion as given in (1.1).

Now consider the question of charges, both of the nucleons and pions. We recall that the neutron and proton can be described as two charge states of a nucleon. The nucleon charge operator is denoted by $\boldsymbol{\tau}$ (bold face quantities denote vectors and operations in charge space), with components:

$$\tau_1 = \begin{pmatrix} 0 & 1 \\ 1 & 0 \end{pmatrix}, \qquad \tau_2 = \begin{pmatrix} 0 & -i \\ i & 0 \end{pmatrix}, \qquad \tau_3 = \begin{pmatrix} 1 & 0 \\ 0 & -1 \end{pmatrix}.$$

The eigenvalues ± 1 of τ_3 describe states of proton and neutron, respectively.

The pions have three charge states ($+e$, 0, $-e$), and so are described by three independent real fields[6] (ϕ_1, ϕ_2, ϕ_3).

The concept of charge independence of the nuclear forces can be built

---

[5] See, for example, Bethe and de Hoffmann, *ibid.*, p. 300.

[6] Recall that ϕ_3 describes the emission and absorption of *neutral* pions, while $\frac{1}{\sqrt{2}}(\phi_1 \pm i\phi_2)$ describe the *charged* pions. For a discussion of the formalism of charged and neutral fields, see G. Wentzel, *Quantum Theory of Fields*, Interscience, New York, 1949.

into the pion-nucleon interaction in a way first suggested by Kemmer.[7] If a charge variable $\boldsymbol{\tau}$ is attributed to the nucleon as well as its spin $\vec{\sigma}$, the "charge symmetric" coupling to the pion field is:

$$(\tau_1\phi_1 + \tau_2\phi_2 + \tau_3\phi_3).$$

If we think of (ϕ_1, ϕ_2, ϕ_3) as the components of a vector $\boldsymbol{\phi}$ in charge space, this can be written as a scalar product:

$$(\tau_1\phi_1 + \tau_2\phi_2 + \tau_3\phi_3) = \boldsymbol{\tau} \cdot \boldsymbol{\phi}.$$

This is a scalar under rotation in charge space and leads to two-nucleon forces with coefficient $\boldsymbol{\tau}_1 \cdot \boldsymbol{\tau}_2$, obviously charge independent.[8]

If the arguments on charge are now combined with the spin arguments we are led to an interaction Hamiltonian density of the form:

$$H_{\text{int}} = F\rho(\vec{r})\,\vec{\sigma} \cdot \vec{\nabla}(\boldsymbol{\tau} \cdot \boldsymbol{\phi}), \tag{1.2}$$

where F is a coupling constant measuring the strength of the interaction. This form of the interaction is known as the "cut-off" or "static" model, and will form the main basis of our discussion of scattering and photoproduction. The presence of the gradient operator implies that the lowest angular momentum involved will be p-wave. We can anticipate then that p-waves will be important in pion-nucleon scattering. The emission and absorption of pions into and from p-states is consistent with the odd parity of the pions, provided the nucleon is held fixed (no recoil approximation).

From a more fundamental point of view it is of interest that essentially (1.2) results from the non-relativistic reduction of the "local γ_5" theory. If we wish to build a linear relativistic theory of pion-nucleon interaction, the fact that the pion is pseudoscalar fixes the form of the interaction, provided derivative couplings are excluded.[9] Thus, the only invariant pseudoscalar bilinear in the nucleon field is $(\bar{\psi}\gamma_5\psi)$, where ψ is a Dirac spinor, $\bar{\psi} = \psi^\dagger\beta$, and $\gamma_5 = \gamma_1\gamma_2\gamma_3\gamma_4$. Consequently the local, relativistic, charge-symmetric interaction is:

$$H_{\text{int}} = -iG\bar{\psi}\gamma_5\boldsymbol{\tau} \cdot \boldsymbol{\phi}\psi, \tag{1.3}$$

where G is a coupling constant.

---

[7] N. Kemmer, *Proc. Cambridge Phil. Soc. 34*, 354 (1938).

[8] $\boldsymbol{\tau}_1 \cdot \boldsymbol{\tau}_2$ has the value -3 for the "isotopic spin" singlet state corresponding to the even (odd) parity, triplet (singlet) spin states of neutron and proton, and has the value $+1$ for the isotopic spin triplet state corresponding to the even (odd) parity, singlet (triplet) spin states of two neutrons, neutron and proton, or two protons.

[9] There is no a priori reason for excluding derivative couplings in a relativistic theory, but one can give various more or less convincing philosophical and aesthetic arguments such as simplicity, renormalizability, and so on. See, for example, S. S. Schweber, H. A. Bethe, F. de Hoffmann, *Mesons and Fields, Vol. I*, Row, Peterson, Evanston, Ill., 1955, p. 332, for one approach to justification.

If one considers matrix elements of (1.3) involving only nonrelativistic nucleons, it is easy to show that one obtains the effective nonrelativistic interaction:

$$H_{\text{int}}^{NR} = \frac{G}{2M} \psi^\dagger \vec{\sigma} \cdot \vec{\nabla}(\boldsymbol{\tau} \cdot \boldsymbol{\phi})\psi, \tag{1.4}$$

where ψ, $\psi^\dagger$ are now Pauli spinors. We note that this is equivalent to (1.2), provided $F = G/2M = .075\ G$.

We remark here that one of the peculiarities of the local γ_5 interaction is the fact that γ_5 connects nucleon-nucleon states rather weakly (note that $F = .075\ G$), but connects nucleon-antinucleon states strongly. This means that virtual pairs of nucleons tend to be formed easily and produce difficult problems in calculation. The "static" theory based on (1.2) avoids the virtual pairs problem by excluding nucleon pairs from the formalism. This certainly destroys the Lorentz invariance of the theory and restricts its application to nonrelativistic motion of the nucleons. But as Chew emphasizes, this is not too serious when we know that at relativistic energies there are other complications such as K-meson interactions to be included in a complete theory. A conceivable drawback of the "static" theory is that it is "non-local." If the nucleon density $\rho(\vec{r})$ is made a delta function, the theory is full of divergences (from the gradient operator). The divergences can be removed only by smearing out the nucleon. It turns out that the details of the smearing are not important; only one parameter is needed to describe it. This parameter is called the "cut-off" energy, ω_{max}. The presence of ω_{max} means that the static theory has two parameters, F and ω_{max}, whereas the local γ_5 theory has only G. The necessity of a smeared out nucleon can be rationalized as a compensation of the ignored effects of nucleon recoil, heavy mesons, etc. The dimensions of the source density $\rho(\vec{r})$ will turn out to be of the order of the *nucleon* Compton wavelength, a plausible size for such compensation.

For future reference we summarize some of the properties of $\rho(\vec{r})$ and of its Fourier transform $v(p^2)$

$$\int \rho(\vec{r})\, d\vec{r} = 1 \tag{1.5}$$

$$v(p^2) = \int e^{-i\vec{p}\cdot\vec{r}} \rho(\vec{r})\, d\vec{r}\quad ; \tag{1.6}$$

v is a function of p^2 only since $\rho(\vec{r})$ is assumed to be spherically symmetric. $v(0) = 1$, and the exact form of v is usually unimportant. The standard choice is:

$$v(p^2) = \begin{cases} 1 \text{ for } p < p_{\text{max}} \\ 0 \text{ for } p > p_{\text{max}} \end{cases}, \tag{1.7}$$

and the *cut-off energy* is defined by $\omega_{max} = \sqrt{p_{max}^2 + 1}$. Other choices of $v(p^2)$ are possible, and some functional form such as a Gaussian might be more plausible. Fortunately the detailed shape of $v(p^2)$, or equivalently $\rho(\vec{r})$, is not significant and (1.7) is a simple, convenient representation.

1.3. Equations of Motion for the Pion Field and Nucleon Variables

In the previous section the significant variables for discussion of the pion-nucleon interaction were found to be the pion fields (ϕ_1, ϕ_2, ϕ_3), expressed as a vector $\boldsymbol{\phi}$ in charge space, and the nucleon spin and charge variables, $\vec{\sigma}$ and $\boldsymbol{\tau}$. At the risk of losing some of the physical content, but with the hope of gaining simplicity, we will treat the pion field as a classical field quantity, but will consider the nucleon variables quantum-mechanically. No important qualitative features of the phenomena are lost by making this semi-classical approximation, although some of the fine points become obscured. We will attempt to indicate the failures of the approximation at the appropriate places in the text.

A field ϕ_i $(i = 1,2,3)$ which quantum-mechanically corresponds to a free spinless particle of mass μ satisfies the Klein-Gordon equation:

$$\Box^2\phi_i - \mu^2\phi_i = 0, \tag{1.8}$$

where $\Box^2$ is the 4-dimensional Laplacian.[10] This equation can be derived from a Lagrangian density in a standard manner.[11] The free field Lagrangian density can be written:

$$\mathscr{L}_{\text{free}} = -\frac{1}{2}\sum_{i=1}^{3}\left[\sum_{\nu=1}^{4}\left(\frac{\partial\phi_i}{\partial x_\nu}\right)^2 + \mu^2\phi_i^2\right], \tag{1.9}$$

and the equation of motion of the field is:

$$\frac{\partial\mathscr{L}}{\partial\phi_i} - \sum_{\nu=1}^{4}\frac{\partial}{\partial x_\nu}\left(\frac{\partial\mathscr{L}}{\partial\left(\dfrac{\partial\phi_i}{\partial x_\nu}\right)}\right) = 0. \tag{1.10}$$

If the fields are in interaction with a source of some sort there will be an interaction Lagrangian density to be added to the free field term. For the static model of Section 1.2 the interaction Lagrangian density is just the negative of the Hamiltonian density in (1.2):

$$\mathscr{L}_{\text{int}} = -\frac{F}{\mu}\rho(\vec{r})\,\vec{\sigma}\cdot\vec{\nabla}(\boldsymbol{\tau}\cdot\boldsymbol{\phi}), \tag{1.11}$$

[10] The factor μ^2 is actually $(\mu c/\hbar)^2$, that is, the square of the reciprocal Compton wavelength of the particle, but in our units $\hbar = c = 1$.

[11] See H. Goldstein, *Classical Mechanics*, Addison-Wesley, Cambridge, 1950, Chapter 11; or G. Wentzel, footnote 6, Chapter 1.

where the factor μ has been inserted to keep F dimensionless. Applying (1.10) to the sum of the free field Lagrangian (1.9) and the interaction term (1.11), one obtains the equation of motion for the field $\boldsymbol{\phi}$ in the presence of a nucleon:

$$(\Box^2 - \mu^2)\boldsymbol{\phi} = -\frac{F}{\mu}\boldsymbol{\tau}\vec{\sigma}\cdot\vec{\nabla}\rho(\vec{r}). \tag{1.12}$$

The equations of motion for the nucleon variables in the presence of the pion field can be most easily obtained by using the standard quantum-mechanical operator equation:

$$\dot{\vec{\sigma}} = \frac{1}{i}[\vec{\sigma}, H] \tag{1.13a}$$

and

$$\dot{\boldsymbol{\tau}} = \frac{1}{i}[\boldsymbol{\tau}, H], \tag{1.13b}$$

where H is the spatial integral of the Hamiltonian interaction density (1.2). Using the commutation properties of $\vec{\sigma}$ and $\boldsymbol{\tau}$ as angular momenta, we find the equations of motion:

$$\dot{\vec{\sigma}} = -2F\int \vec{\sigma}\times\vec{\nabla}(\boldsymbol{\tau}\cdot\boldsymbol{\phi})\rho(\vec{r})\,d\vec{r} \tag{1.14}$$

and

$$\dot{\boldsymbol{\tau}} = -2F\int \boldsymbol{\tau}\times(\vec{\sigma}\cdot\vec{\nabla}\boldsymbol{\phi})\rho(\vec{r})\,d\vec{r}. \tag{1.15}$$

The three coupled equations (1.12), (1.14), (1.15) determine the behavior of the pion-nucleon system.

A simple illustration of the solution of (1.12) is afforded by assuming that the nucleon variables are fixed and constant in time and that the source density is a delta function at the origin. Then the static solution for the pion field is easily shown to be:

$$\boldsymbol{\phi} = \frac{F}{4\pi}\boldsymbol{\tau}\vec{\sigma}\cdot\vec{\nabla}\left(\frac{e^{-r}}{r}\right), \tag{1.16}$$

where μ has been set equal to unity in our units. This static field has the characteristic feature of extending out from the source to distances of the order of the pion Compton wavelength (1.41×10^{-13} cm). It can be thought of as representing a cloud of virtual pions around the nucleon.

The static model can be treated completely quantum-mechanically.[12] For this purpose the interaction Hamiltonian is expressed in terms of creation and

[12] For an excellent discussion of the static model from a modern field theoretical point of view, see G. C. Wick, *Rev. Mod. Phys.* *27*, 339 (1955).

destruction operators. If the usual expansion of the meson field is performed,

$$\phi_j = \sum_q \frac{1}{\sqrt{2\omega_q}} (a_{jq}^\dagger e^{-i\vec{q}\cdot\vec{r}} + a_{jq} e^{i\vec{q}\cdot\vec{r}}), \tag{1.17}$$

the interaction energy (the integral of (1.2) over space) can be written:

$$H_{\text{int}} = \sum_{jq} (V_{jq} a_{jq} + V_{jq}^\dagger a_{jq}^\dagger), \tag{1.18}$$

where

$$V_{jq} = iFv(q^2)\tau_j \frac{\vec{\sigma}\cdot\vec{q}}{\sqrt{2\omega_q}}, \tag{1.19}$$

and the operators $a_{jq}^\dagger$ and a_{jq} are the usual creation and destruction operators for pions with momentum $\vec{q}$ and charge index j. The unperturbed energy is just that of the free pions, since we assume that the nucleon is fixed at rest:

$$H_0 = \sum_{jq} a_{jq}^\dagger a_{jq} \omega_q. \tag{1.20}$$

The scattering and other processes are described in terms of the matrix elements V_{jq}.

CHAPTER 2

Scattering of Pions by Nucleons

The scattering of pions by single nucleons is one important source of information about the interaction between pions and nucleons. With present day accelerators pion beams are available for detailed differential scattering experiments up to energies of several hundred Mev, and for total cross section measurements up to several Bev. Above 300 or 400 Mev there is enough energy available in the center of momentum system of the pion and nucleon to produce additional pions in considerable abundance. Consequently, at high energies inelastic processes are important and it is not easy to interpret the data in a simple way. Below 300 Mev inelastic reactions are sufficiently improbable, if not energetically forbidden, that the scattering can be treated as completely elastic, and a relatively straightforward interpretation can be given.

The usual analysis of scattering is in terms of phase shifts appropriate to an expansion in spherical harmonics, where one keeps only the first few terms in the expansion at low energies. The pion-nucleon scattering can be handled in exactly this way, but because of the concept of charge independence it is useful to consider a separation in charge space as well.

2.1. Isotopic Spin Analysis

We recall that the proton and neutron can be thought of as two charge states of a nucleon, and that the charge or isotopic spin variable is described quantum-mechanically by the operator $\frac{1}{2}\boldsymbol{\tau}$. The factor $\frac{1}{2}$ is present so that the charge operator satisfies the usual commutation rules of angular momentum, just as for the nucleon spin $\frac{1}{2}\vec{\sigma}$. The pion field is described in charge space by the vector $\boldsymbol{\phi}$, and its charge operator can be shown[13] to be:

$$\boldsymbol{t} = \int \boldsymbol{\phi} \times \boldsymbol{\pi} d\vec{r} \quad , \tag{2.1}$$

where $\boldsymbol{\pi} = \dot{\boldsymbol{\phi}}$ is the field momentum canonically conjugate to $\boldsymbol{\phi}$.

Just as with ordinary angular momentum, we can combine the "isotopic spin" of the nucleon with that of the pion field to get the total "isotopic spin" of the pion-nucleon system:

$$\boldsymbol{T} = \tfrac{1}{2}\boldsymbol{\tau} + \boldsymbol{t} \tag{2.2}$$

[13] See G. Wentzel, footnote 6, Sec. 8, or Bethe and de Hoffmann, footnote 4, p. 54.

With the charge symmetric interaction, T is a constant of the motion. To show that this is true, one merely takes the time derivative of (2.2);

$$\dot{T} = \tfrac{1}{2}\dot{\boldsymbol{\tau}} + \int (\dot{\boldsymbol{\phi}} \times \boldsymbol{\pi} + \boldsymbol{\phi} \times \dot{\boldsymbol{\pi}})\, d\vec{r}. \tag{2.3}$$

The second term on the right hand side can be shown to be the negative of the first. Thus, one notes that the first term under the integral sign vanishes from the definition of $\boldsymbol{\pi}$, while the second part of the integrand can be simplified by use of (1.12):

$$\boldsymbol{\phi} \times \dot{\boldsymbol{\pi}} = \boldsymbol{\phi} \times \ddot{\boldsymbol{\phi}} = \boldsymbol{\phi} \times (\nabla^2 - 1)\boldsymbol{\phi} + \boldsymbol{\phi} \times (F\boldsymbol{\tau}\vec{\sigma} \cdot \vec{\nabla}\rho)$$

or

$$\boldsymbol{\phi} \times \dot{\boldsymbol{\pi}} = -F(\boldsymbol{\tau} \times \boldsymbol{\phi})\vec{\sigma} \cdot \vec{\nabla}\rho + \boldsymbol{\phi} \times \nabla^2\boldsymbol{\phi}. \tag{2.4}$$

Substitution of (2.4) into (2.3) and integration by parts leads to an integral which is identifiable from (1.15) as $(-\tfrac{1}{2}\dot{\boldsymbol{\tau}})$, so that

$$\dot{T} = 0. \tag{2.5}$$

This result also holds in a rigorously quantum-mechanical treatment.

For a single pion the eigenvalues of t_3 are $(+1,0,-1)$ corresponding to the eigenvalue $t = 1$ for the total pion isotopic spin. The combination of $\tfrac{1}{2}\tau = \tfrac{1}{2}$ and $t = 1$ leads to the two possible states:

$$T = \tfrac{1}{2} \quad \text{and} \quad T = \tfrac{3}{2}.$$

If we consider the scattering of various combinations of charges for the pions and nucleons, we get different mixtures of these two states. Let f_{2T} be the scattering amplitude for the state with total isotopic spin T. Then in $(\pi^+ + p)$ scattering, for example, only the amplitude f_3 enters (the operator T_3 has eigenvalue $+\tfrac{3}{2}$). For other scattering processes one can use the usual vector addition coefficients to obtain the results:

$$\pi^+ + p \rightarrow \pi^+ + p \qquad f = f_3 \tag{2.6}$$

$$\pi^- + p \rightarrow \pi^- + p \qquad f = \tfrac{1}{3}(2f_1 + f_3) \tag{2.7}$$

$$\pi^- + p \rightarrow \pi^0 + n \qquad f = \frac{\sqrt{2}}{3}(-f_1 + f_3) \tag{2.8}$$

$$\pi^0 + p \rightarrow \pi^0 + p \qquad f = \tfrac{1}{3}(f_1 + 2f_3). \tag{2.9}$$

The first three processes are the only ones that are studied directly, but the data can then be used to extract f_1 and f_3 separately and so calculate the cross section for (2.9). There are the same number of mirror scattering processes for pions on neutrons, with the same amplitudes as in (2.6) to (2.9).

2.2. Essential Features of the Scattering up to 400 Mev

The total cross sections for positive and negative pions on protons are shown in Fig. 1 for pion kinetic energies up to 2 Bev. In the low-energy region where the scattering is elastic the π^+ curve corresponds to process

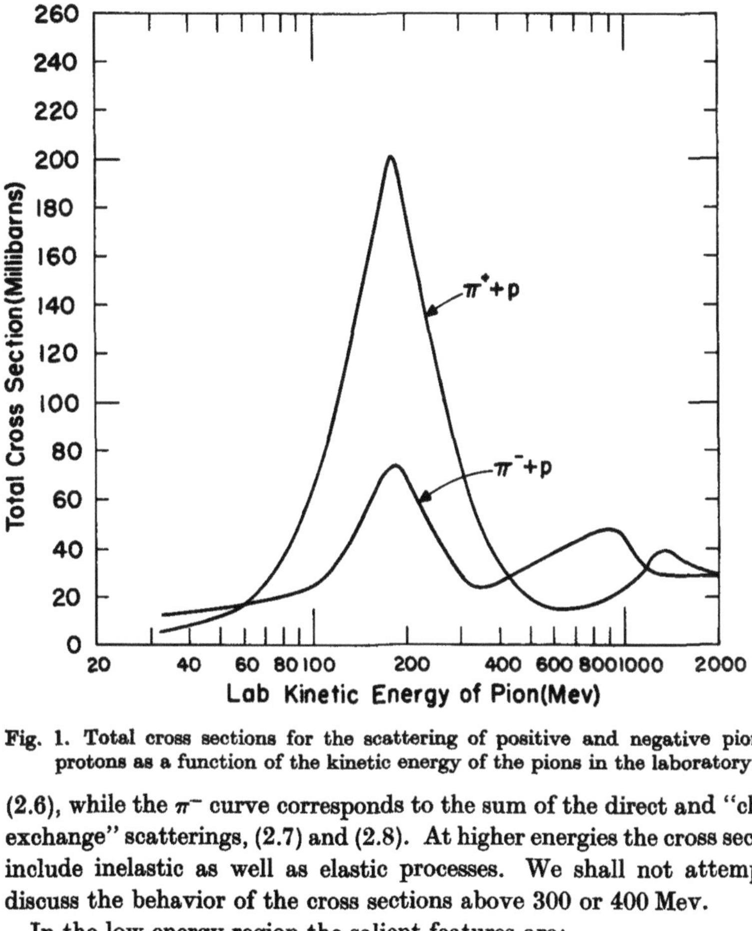

Fig. 1. Total cross sections for the scattering of positive and negative pions by protons as a function of the kinetic energy of the pions in the laboratory.

(2.6), while the π^- curve corresponds to the sum of the direct and "charge exchange" scatterings, (2.7) and (2.8). At higher energies the cross sections include inelastic as well as elastic processes. We shall not attempt to discuss the behavior of the cross sections above 300 or 400 Mev.

In the low-energy region the salient features are:

(1) The $(\pi^+ + p)$ cross section rises very rapidly with increasing pion energy in the energy range up to 200 Mev (lab kinetic energy).

(2) From about 100 Mev to 300 Mev the $(\pi^- + p)$ total cross section is very closely equal to one third of the $(\pi^+ + p)$ total cross section, and in fact the three processes (2.6), (2.7), (2.8) are in roughly constant ratio of 9 : 1 : 2.

(3) The cross sections show a pronounced resonance behavior in the neighborhood of 190 Mev with a width of about 140 Mev.

The rapid rise of $(\pi^+ + p)$ at low energies is roughly as q^4 in the energy range from 30 to 120 Mev. It is consistent with the perturbation result based on (1.2) or (1.18):

$$\sigma = 16\pi f^4 \frac{q^4}{1+q^2} \tag{2.10}$$

where $f^2 = F^2/4\pi$, and the cross section is in units of $\mu^{-2} \simeq 20$ mb. This emphasizes the importance of p-wave scattering (the lth partial cross section is proportional to q^{4l} at low energies). The perturbation result is certainly not valid, however, since it predicts equality for the cross sections (2.6) and (2.7) and a cross section of $\frac{4}{9}\sigma$ for the exchange scattering (2.8).

Inspection of the expressions (2.6) to (2.8) for the scattering amplitudes in terms of isotopic spin states shows that the approximate ratio 9 : 1 : 2 independent of energy can be explained by the assumption:

$$f_3 \gg f_1.$$

This shows that the $T = \frac{3}{2}$ state is much more important than the $T = \frac{1}{2}$ state, at least for pion energies from 100 to 300 Mev. The resonance behavior can now be ascribed to a resonance in the $T = \frac{3}{2}$ state (some phase shift approaches or goes through 90°).

The magnitude of the cross section for $(\pi^+ + p)$ at its peak gives a clue as to the important angular momentum. Assuming that it is a true resonance for a single angular momentum, we recall that for the scattering of unpolarized particles the maximum value of the cross section is:

$$\sigma_{\text{max}} = 4\pi\lambda\!\!\!^{-2} \frac{(2j+1)}{(2s_1+1)(2s_2+1)}, \tag{2.11}$$

where s_1, s_2 are the spins of the colliding particles, j is the angular momentum of the resonance, and $\lambda\!\!\!^{-}$ is the reduced wavelength at resonance. The observed peak cross section is about 200 mb, while $\pi\lambda\!\!\!^{-2}$ at 190 Mev is about 24 mb. This implies that $(2j+1) \lesssim 4$, or $j = \frac{3}{2}$. (The value of $j = \frac{1}{2}$ can be excluded on the grounds that half of the cross section at the peak would then have to be nonresonant scattering).[14] This is the famous (3,3) resonance[15] which governs the behavior of all low-energy pion phenomena. It is most logically attributed to the state with $l = 1$, and so is a $P_{3/2}$ resonance.[16]

[14] The arguments about the resonance can incidentally be used to confirm the assignment of zero spin for the pion. If the pion spin is not zero, it would most likely be $s_1 = 2$ since the decay of the neutral pion requires even values. With spin two, the condition on the angular momentum of the resonant state would be $(2j+1) \lesssim 20$, or $j \simeq \frac{19}{2}$. Such a large angular momentum at the observed energy seems very unreasonable.

[15] The notation (3,3) corresponds to $(2T, 2j)$.

[16] A simple description of the pion-nucleon scattering and photoproduction in terms of a single level resonance formula is remarkably successful (M. Gell-Mann and K. M. Watson, *Annual Reviews of Nuclear Science 4*, 219 (1954)).

The angular distributions of the scattering provide additional valuable information. Just the qualitative behavior of the angular distributions confirms the idea that a true resonance occurs near 190 Mev. Fig. 2

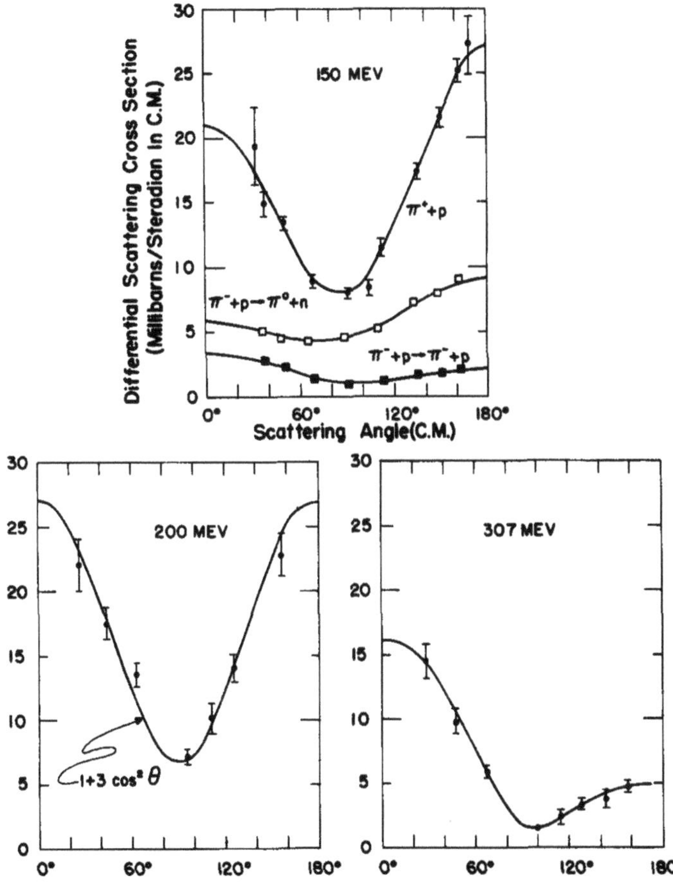

Fig. 2. Differential scattering cross sections of pions by protons in the center of momentum coordinate system. The upper diagram shows data at 150 Mev laboratory kinetic energy on the scattering of positive pions by protons, and both the direct and charge exchange scattering of negative pions by protons. The lower diagrams show data for positive pions only at 200 and 307 Mev (see footnote 17). At 200 Mev, the curve shown is proportional to $(1 + 3\cos^2\theta)$, while at the other energies the curves are merely empirical fits to the experimental points.

shows a typical set of angular distribution measurements at three energies: one below the peak cross section, one very near it, and one well above the peak.[17] The 150 Mev results include differential cross sections for positive

[17] The 150 Mev data in Fig. 2 are those of Ashkin, Blaser, Feiner, and Stern (*Phys. Rev. 101*, 1149 (1956)), while the 200 and 307 Mev data are those of Mukhin, Ozerov, and Pontecorvo (*Soviet Physics JETP 4*, 237 (1957)).

pions on protons and negative pions on protons, both direct and charge exchange. The 200 Mev and 307 Mev results are for positive pions only. Qualitatively the symmetry of the angular distribution about 90° near the peak energy, and the skewness (one way below and the other way above that energy) strongly imply the presence of an actual resonance in a manner very familiar from low-energy nuclear reactions. The detailed analysis of the angular distributions will be discussed in the next section.

2.3. Angular Distributions and Phase Shifts

The angular distributions for the various scatterings can be analyzed for scattering phase shifts. To do this, one needs to expand the scattering amplitudes f_1 and f_3 into partial waves. For energies up to and including the resonance it is sufficient to include only $\ell = 0$ and $\ell = 1$ waves. Then there are six phase shifts corresponding to the six states specified by the quantum numbers T, ℓ, and j as shown in Table 2, where for the s-waves,

Table 2. Scattering Phase Shifts

	$\ell = 0$	$\ell = 1$	
	$j = \frac{1}{2}$	$j = \frac{1}{2}$	$j = \frac{3}{2}$
$T = \frac{1}{2}$	δ_1	δ_{11}	δ_{13}
$T = \frac{3}{2}$	δ_3	δ_{31}	δ_{33}

the subscript is $2T$, and for the p-waves, the subscripts are $2T$, $2j$. Each phase shift is connected to a scattering amplitude for its particular state in the standard manner:

$$a = \frac{1}{2iq}(e^{2i\delta} - 1), \tag{2.12}$$

where q is the relative wave number in the center of momentum coordinate system.

To relate these amplitudes to the angular distribution we first write the total scattering amplitude (for a given isotopic spin state) in the form:

$$f = A + B\vec{q}' \cdot \vec{q} + iC\vec{\sigma} \cdot (\vec{q}' \times \vec{q}), \tag{2.13}$$

where $\vec{q}$, $\vec{q}'$ are momentum vectors of length q in the directions of the incident and scattered pions in the center of mass. The form (2.13) is the most general that can be written containing s- and p-waves only. A corresponds to s-wave scattering, while B and C describe the direct and "spin flip" p-wave scattering. All three coefficients are functions of the pion energy. The form (2.13) is understood as an operator taken between initial and final nucleon spin states. A cross section for an unpolarized target is given by:

$$\frac{d\sigma}{d\Omega} = \tfrac{1}{2} \sum_{s_i, s_f} |(s_f|f|s_i)|^2 . \tag{2.14}$$

Performing this operation on (2.13) yields the result:

$$\frac{d\sigma}{d\Omega} = |A + Bq^2 \cos\theta|^2 + |Cq^2|^2 \sin^2\theta. \tag{2.15}$$

For the various scatterings (2.6) to (2.9) we merely replace each of the amplitudes, A, B, C by the appropriate sum of amplitudes for $T = \frac{1}{2}$ and $T = \frac{3}{2}$.

We must now connect the coefficients A, B, C to the amplitudes (2.12) for the states of definite parity and angular momentum. First define the operators:

$$\begin{aligned} J_{\frac{1}{2}}(\vec{q}\,', \vec{q}) &= \vec{q}\,' \cdot \vec{q} + i\vec{\sigma} \cdot (\vec{q}\,' \times \vec{q}) \\ J_{\frac{3}{2}}(\vec{q}\,', \vec{q}) &= 2\vec{q}\,' \cdot \vec{q} - i\vec{\sigma} \cdot (\vec{q}\,' \times \vec{q}). \end{aligned} \tag{2.16}$$

These operators are p-wave projection operators for the angular momentum states $j = \frac{1}{2}$ and $j = \frac{3}{2}$ for the scattered pion, in the sense that, given the state $\Psi_0 = \alpha q$ describing a nucleon with spin up and a p-wave pion incident along the z-axis,

$$\begin{aligned} J_{\frac{1}{2}}\Psi_0 &= q^2\Psi_{\frac{1}{2}}^{\frac{1}{2}}(q') \\ J_{\frac{3}{2}}\Psi_0 &= q^2\Psi_{\frac{3}{2}}^{\frac{1}{2}}(q'), \end{aligned}$$

where $\Psi_j^m(q')$ is an eigenstate of total angular momentum j of the scattered pion $\vec{q}\,'$. (These functions are not normalized in the usual sense, but have the appropriate normalization for the scattering problem, e.g. $\Psi_{\frac{3}{2}}^{\frac{1}{2}}(q') = q_3'\alpha + (q_1' + iq_2')\beta$). With these projection operators we can rewrite (2.13) in the form:

$$f_{2T} = a_{2T} + a_{2T,1}\frac{J_{\frac{1}{2}}(\vec{q}\,', \vec{q})}{q^2} + a_{2T,3}\frac{J_{\frac{3}{2}}(\vec{q}\,', \vec{q})}{q^2}, \tag{2.17}$$

where the connection with the coefficients A, B, C is:

$$\begin{aligned} A_{2T} &= a_{2T} \\ B_{2T}q^2 &= a_{2T,1} + 2a_{2T,3} \\ C_{2T}q^2 &= a_{2T,1} - a_{2T,3} . \end{aligned} \tag{2.18}$$

With these connection formulas and the definition of a in terms of the phase shift (2.12), the differential cross section (2.15) can be written entirely in terms of the six phase shifts. For example, the *total* cross section for $(\pi^+ + p)$ scattering is:

$$\sigma_{\pi^+ + p} = \frac{4\pi}{q^2} (\sin^2 \delta_3 + \sin^2 \delta_{31} + 2 \sin^2 \delta_{33}). \tag{2.19}$$

We see here explicitly the maximum value $8\pi/q^2$ when $\delta_{33} = \pi/2$ and the other phase shifts are small. If all phase shifts are neglected except that for the (3,3) state, then the *differential* cross section for $\pi^+ + p$ takes the form:

$$\frac{d\sigma}{d\Omega} = \frac{\sin^2 \delta_{33}}{q^2} (1 + 3 \cos^2 \theta). \tag{2.20}$$

From Fig. 2 it can be seen that near the peak of the total cross section curve (~200 Mev) this angular dependence is a good fit to the data.

There have been a great number of experiments on the angular distributions at various energies up to 300 Mev, of which those shown in Fig. 2 are typical. The phase shifts are not uniquely determined by the data, and there have been two sets of phase shifts discussed: the "Fermi" solution and the "Yang" solution which differs from the Fermi solution by giving the opposite sign to the spin-flip amplitude C in (2.13). The Fermi solution is characterized by a large positive value of δ_{33} and $\delta_{31} \simeq 0$. The Yang solution has $\delta_{31} > \delta_{33}$. Davidon and Goldberger[18] have used the spin-flip dispersion relations to exclude the Yang solution.

Below 100 Mev the experiments are fitted well by the following phase shifts:

$$\begin{aligned} \delta_1 &= 0.17q &&= 9.7°q \\ \delta_3 &= -0.11q &&= -6.3°q \\ \delta_{33} &= 0.235q^3 &&= 13.5°q^3, \end{aligned} \tag{2.21}$$

with $\delta_{11} \simeq \delta_{13} \simeq \delta_{31} \simeq 0$. The momentum dependence is appropriate to to the orbital angular momentum involved ($\delta_\ell \sim q^{2\ell+1}$).

Above 100 Mev, δ_{33} rises more rapidly than q^3, passing through 90° at about 195 Mev ($q = 1.65$, $\omega_q = 1.93$). The other phase shifts are not at all well determined, but are small—generally much less than 20°. There is evidence that δ_{31} is negative at all energies. It will be seen later that in the cut-off theory a large positive 3,3 phase shift is expected, while all the other p-wave phases are small and negative.

The s-wave phase shifts seem to follow the linear dependence on momentum of (2.21) up to at least 350 Mev, although at that energy d-wave phase shifts are becoming significant (~10°) and the situation is not clear cut.

---

[18] W. C. Davidon and M. L. Goldberger, *Phys. Rev.* *104*, 1119 (1956). Dispersion relations are discussed in Sec. 3.3.

CHAPTER 3

Theoretical Treatment of the Scattering—Effective Range Approximation and the Dispersion Relations

The static model defined by the Hamiltonian (1.2) is a model describing only p-wave scattering. No s-wave scattering is predicted by such an interaction.[19] Fortunately in the energy range up to and including the resonance, the s-wave phase shifts given by (2.21) are less than 20° and are small compared to the dominant 3,3 phase shift. Thus a model which ignores s-wave effects will still be able to give a reasonably satisfactory description of the pion-nucleon interaction except at very low energies.

A proper quantum-mechanical treatment of the static model is given in the references cited in footnotes 1, 2, and 3. Here we will follow a semi-classical approach based on the equations of Section 1.3. After a discussion of the effective range approximation in these semi-classical terms, the correlation of the experimental data by means of exact dispersion relations will be described.

3.1. Effective Range Approximation

In Section 1.3 equations of motion for the static model were obtained with the approximation that the nucleon spin and charge variables were quantum-mechanical operators, but that the pion field was a classical variable. The equation for the pion field was (1.12), while the equations for the spin and charge variables were (1.14) and (1.15). These equations will be solved in the following approximate way[20]: The first order solution of the equations for $\vec{\sigma}$ and $\boldsymbol{\tau}$ will be used to give an approximate potential acting on the pion field $\boldsymbol{\phi}$. Then the wave equation for the field will be solved essentially exactly to yield explicit expressions for the various p-wave phase shifts as functions of energy.

We are interested in a scattering problem in which a pion of momentum $\vec{q}$ and energy ω is scattered by a nucleon. The pion field will therefore have the time dependence exp $(-i\omega t)$, and it will be assumed that the

[19] For a discussion of a generalized static model which includes both s- and p-waves, see Drell, Friedman, and Zachariasen, *Phys. Rev.* *104*, 236 (1956).

[20] The method is similar to that used by S. F. Edwards and P. T. Matthews, *Phil. Mag.* *2*, 176 (1957).

time varying parts of $\vec{\sigma}$ and $\boldsymbol{\tau}$ have similar time dependences. The field equation (1.12) then becomes:

$$(\nabla^2 + q^2)\boldsymbol{\phi} = -F\boldsymbol{\tau}\vec{\sigma} \cdot \vec{\nabla}\rho(\vec{r}). \tag{3.1}$$

The equations (1.14) and (1.15) for $\vec{\sigma}$ and $\boldsymbol{\tau}$ are now solved approximately by assuming that $\boldsymbol{\tau}$ and $\vec{\sigma}$ are essentially constant on the right-hand sides of these equations. With

$$\vec{\sigma}(t) = \vec{\sigma} + \vec{\sigma}_1 e^{-i\omega t}$$

$$\boldsymbol{\tau}(t) = \boldsymbol{\tau} + \boldsymbol{\tau}_1 e^{-i\omega t}, \tag{3.2}$$

where $\vec{\sigma}_1$ and $\boldsymbol{\tau}_1$ are assumed to be small, time-dependent terms, one finds

$$\vec{\sigma}_1 = \frac{2F}{i\omega} \vec{\sigma} \times \int \vec{\nabla}(\boldsymbol{\tau} \cdot \boldsymbol{\phi})\rho \, d\vec{r}, \tag{3.3}$$

$$\boldsymbol{\tau}_1 = \frac{2F}{i\omega} \boldsymbol{\tau} \times \int (\vec{\sigma} \cdot \vec{\nabla} \boldsymbol{\phi})\rho \, d\vec{r}. \tag{3.4}$$

In the equation (3.1) for $\boldsymbol{\phi}$ we are interested in terms proportional to $\exp(-i\omega t)$. Thus the relevant part of $\boldsymbol{\tau}\vec{\sigma}$ is:

$$(\boldsymbol{\tau}\vec{\sigma})_{\text{eff}} = \vec{\sigma}\boldsymbol{\tau}_1 + \boldsymbol{\tau}\vec{\sigma}_1.$$

The ambiguity about the ordering of noncommuting operators in this expression is avoided by symmetrizing it in the usual way:

$$(\boldsymbol{\tau}\vec{\sigma})_{\text{eff}} = \tfrac{1}{2}[(\vec{\sigma}\boldsymbol{\tau}_1 + \boldsymbol{\tau}_1\vec{\sigma}) + (\boldsymbol{\tau}\vec{\sigma}_1 + \vec{\sigma}_1\boldsymbol{\tau})].$$

With this form one obtains the following wave equation for $\boldsymbol{\phi}$:

$$(\nabla^2 + q^2)\boldsymbol{\phi} = -\frac{2F^2}{i\omega}(\vec{\nabla}\rho) \cdot \int [\vec{\nabla}'(\boldsymbol{\tau} \times \boldsymbol{\phi}) + \vec{\sigma} \times \vec{\nabla}'\boldsymbol{\phi}]\rho(\vec{r}')\, d\vec{r}'. \tag{3.5}$$

Equation (3.5) can be thought of as a wave equation for $\boldsymbol{\phi}$ with the right-hand side involving a non-local potential operator. It looks rather complicated but is relatively simple. We recall that the analysis of the scattering was aided by a decomposition into states of definite (l, j, T). The gradient operator on the right-hand side implies that only p-waves are present. First consider the separation into states of definite total angular momentum. If the field is factored into radial and angle-spin parts:

$$\boldsymbol{\phi} = U_j(r)\Psi_j^m/r,$$

where Ψ_j^m is an eigenstate of $j = \ell \pm \frac{1}{2}$ and $j_z = m$, then substitution into (3.5) leads to the p-wave equation:

$$\left(\frac{d^2}{dr^2} - \frac{2}{r^2} + q^2\right)U_j = -\frac{2F^2}{3\omega} r \frac{d\rho}{dr} 4\pi \int_0^\infty r' \frac{d\rho}{dr'} [\xi_j U_j + i\tau \times U_j]\, dr',$$

where

$$\xi_j = +1 \quad \text{for} \quad j = \tfrac{3}{2}, \quad \text{and} \quad \xi_j = -2 \quad \text{for} \quad j = \tfrac{1}{2}.$$

A completely analogous separation into states of total isotopic spin leads to the result:

$$\left(\frac{d^2}{dr^2} - \frac{2}{r^2} + q^2\right)U_{T,j} = -\frac{2F^2}{3\omega} 4\pi(\xi_j + \xi_T) r \frac{d\rho}{dr} \int_0^\infty r' \frac{d\rho}{dr'} U_{T,j}\, dr'.$$

Introducing the unrationalized coupling constant $f^2 = F^2/4\pi$ and the notation $\alpha = (2T,2j)$, with $\alpha = 1$ meaning (1,1), $\alpha = 2$ meaning $(1,3) = (3,1)$ and $\alpha = 3$ (3,3), this equation becomes:

$$\left(\frac{d^2}{dr^2} - \frac{2}{r^2} + q^2\right)U_\alpha = -\frac{\lambda_\alpha}{\omega} (4\pi)^2 r \frac{d\rho}{dr} \int_0^\infty r' \frac{d\rho}{dr'} U_\alpha(r')\, dr', \tag{3.6}$$

where

$$\lambda_\alpha = \frac{f^2}{3} \begin{pmatrix} -8 \\ -2 \\ +4 \end{pmatrix}. \tag{3.7}$$

We now consider the effective range approach to the solutions of (3.6), in complete analogy with the derivation used in nucleon-nucleon scattering.[21] First it should be noted that outside the range of $\rho(r)$, $U(r)$ will be a linear combination of the solutions of (3.6) with the right-hand side put equal to zero. This asymptotic form is denoted by $W(r)$:

$$W(r) = q^2 r[\cot \delta\, j_1(qr) - n_1(qr)], \tag{3.8}$$

where $j_1(x)$ and $n_1(x)$ are spherical Bessel functions of order unity, and δ is the phase shift $\left(\text{outside the centrifugal barrier, } W \sim \sin\left(qr - \frac{\pi}{2} + \delta\right)\right)$. The normalization of $W(r)$ is chosen for later convenience. If one now writes down (3.6) for two energies ω_a and ω_b and proceeds in the standard manner, one obtains the result:

$$q_b^3 \cot \delta_b - q_a^3 \cot \delta_a = (\omega_b^2 - \omega_a^2) \int_0^\infty \left(W_a W_b - \frac{1}{r^2} - U_a U_b\right) dr - \lambda I_a I_b \left(\frac{1}{\omega_b} - \frac{1}{\omega_a}\right), \tag{3.9}$$

[21] H. A. Bethe, *Phys. Rev. 76*, 38 (1949).

where

$$I = 4\pi \int_0^\infty r \frac{d\rho}{dr} U(r)\, dr. \tag{3.10}$$

The added term involving the I's comes from the right-hand side of (3.6). To evaluate I we need to consider the solution of (3.6) at large distances and compare it with $W(r)$ given by (3.8). With a Green's function appropriate to the left-hand side of (3.6), $U(r)$ is given by:

$$U(r) = q^2 r\left[\cot\delta\, j_1(qr) - \frac{4\pi\lambda I}{\omega q}\int_0^\infty r'^2 \frac{d\rho}{dr'} j_1(qr_<) n_1(qr_>)\, dr'\right].$$

At distances outside $\rho(r)$ this becomes:

$$U \to q^2 r\left[\cot\delta\, j_1(qr) - n_1(qr)\frac{4\pi\lambda I}{\omega q}\int_0^\infty r'^2 \frac{d\rho}{dr'} j_1(qr')\, dr'\right].$$

Comparison with (3.8) for $W(r)$ shows that:

$$1 = \frac{4\pi\lambda I}{\omega q}\int_0^\infty r'^2 \frac{d\rho}{dr'} j_1(qr')\, dr'. \tag{3.11}$$

To get an explicit form for I, we assume that $\rho(r')$ is small in extent compared to the wavelength q^{-1}. Then we can expand the spherical Bessel function $j_1(qr) \simeq qr/3$, and obtain finally

$$I = -\frac{\omega}{\lambda}. \tag{3.12}$$

Insertion of this value of I into (3.9) leads to the formula:

$$q_b^3 \cot\delta_b - q_a^3 \cot\delta_a = \frac{1}{\lambda}(\omega_b - \omega_a) + (\omega_b^2 - \omega_a^2)\int_0^\infty \left(W_a W_b - \frac{1}{r^2} - U_a U_b\right) dr$$

as the almost exact relation between phase shifts at two energies. The effective range approximation is obtained by assuming that the effective range integral is essentially constant.

The standard form of the effective range approximation is obtained by choosing $\omega_a = 0$. For $\omega_a = 0$, $q_a = i$ and $\cot\delta_a = i$ (to prevent an exponentially growing solution at infinity). Then the formula takes on the form:

$$\frac{q^3 \cot\delta_\alpha}{\omega} - \frac{1}{\omega} = \frac{1}{\lambda_\alpha}(1 - \omega r_\alpha), \tag{3.13}$$

where the subscript b has been dropped and the subscript α restored. The "effective range" r_α is defined by:

$$r_\alpha = \lambda_\alpha \int_0^\infty \left(U_0^2 + \frac{1}{r^2} - W_0^2\right) dr.$$

For a compact source distribution one can show that to a good approximation:

$$r_\alpha = \frac{8\lambda_\alpha}{\pi^2} \int_0^\infty |v(k)|^2 \, dk, \tag{3.14}$$

where $v(k)$ is the Fourier transform of the source density, (1.6).[22]

The effective range formula (3.13) was first obtained by Chew and Low (footnote 2). It is usually applied with the $1/\omega$ term omitted (the form (3.13) is sometimes called the Serber modification), and with ω replaced by $\omega^* = \omega_q + q^2/2M$ to include kinematic effects. We note first of all that (3.13) predicts that $\delta_{33}(\alpha = 3)$ will be positive and the other p-wave phase shifts will be negative, according to the sign of λ_α. Similarly the result for r_α shows that the effective range for the 3,3 state will be positive, while the others are negative. This means that δ_{33} will be positive and show a resonance at $\omega^* = 1/r_3$, while all the other p-wave phase shifts will be negative and nonresonant. This is in good qualitative agreement with the observations.

For comparison with (3.13), we note that the Born approximation yields phase shifts

$$\delta_{\alpha\text{Born}} = \frac{q^3\lambda_\alpha}{\omega},$$

which give a cotangent formula:

$$\frac{q^3 \cot \delta_{\alpha\text{Born}}}{\omega} \simeq \frac{1}{\lambda_\alpha}. \tag{3.15}$$

Comparison with (3.13) shows that for negative effective ranges ($\alpha = 1,2$) the Born approximation is qualitatively correct, but for positive effective range ($\alpha = 3$) the Born result is grossly inadequate, except at very low energies.

According to the effective range formula, a plot of $q^3 \cot \delta/\omega^*$ for the 3,3 phase shift should yield a straight line with an extrapolated intercept

---

[22] It is of interest to note that the effective range r_α defined by (3.14) is closely related to the "spin inertia"

$$\frac{1}{a} = \frac{2}{\pi} \int_0^\infty |v(k)|^2 dk = \iint \frac{\rho(\vec{r})\, \rho(\vec{r}')}{|\vec{r} - \vec{r}'|} \, d\vec{r}\, d\vec{r}'$$

introduced by W. Pauli, *Meson Theory of Nuclear Forces*, Interscience, New York, 1946, p. 17.

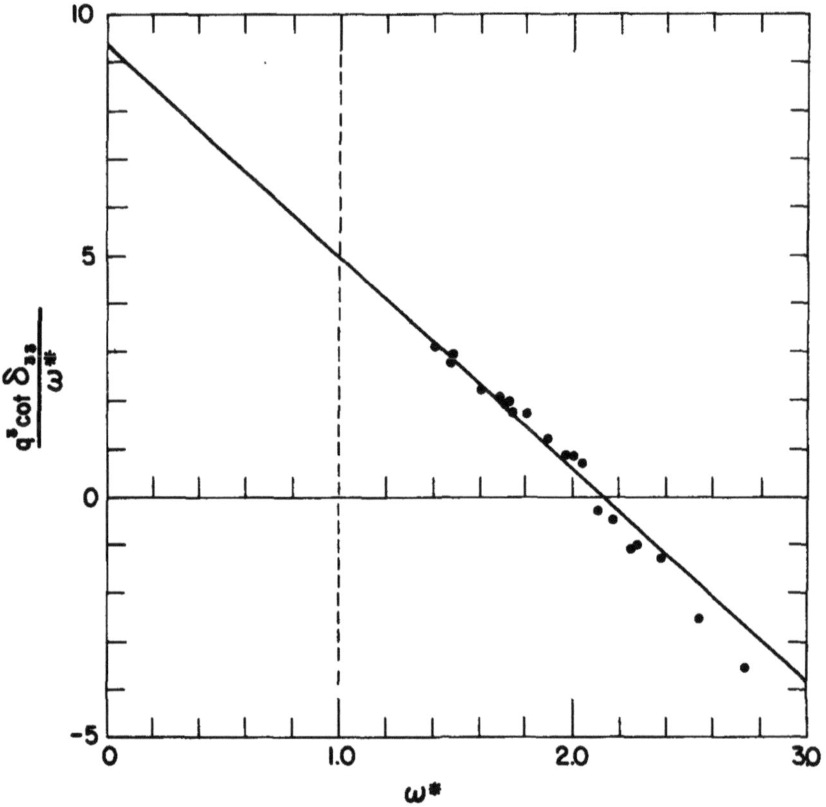

Fig. 3. Effective range plot for the 3,3 phase shift. $q^3 \cot \delta_{33}/\omega^*$ is plotted as a function of $\omega^* = \omega_q + q^2/2M$. The intercept at $\omega^* = 0$ implies a coupling constant $f^2 = 0.08 \pm 0.02$.

at $\omega^* = 0$ of λ_3^{-1}. Such a plot is shown in Fig. 3. The intercept at $\omega^* = 0$ implies a coupling constant $f^2 = 0.08 \pm 0.02$. From the position of the resonance at $\omega_0^* = 2.1$ we can infer the value of the other theoretical parameter, the cut-off ($\omega_{\max}$). With the square distribution (1.7) for $v(k)$ in (3.14), the relation is approximately:

$$r_{33} \simeq f^2 \omega_{\max}. \tag{3.16}$$

The observed values of f^2 and ω_0^* imply $\omega_{\max} \simeq 6$. The large value of $\omega_{\max}$ shows that the source density is only spread out over about a nucleon Compton wavelength. This is quite a compact source distribution and justifies the approximations made in solving (3.6), provided we do not consider energies far above the resonance.

Recently Puppi and Stanghellini[23] have analyzed the latest data in

[23] G. Puppi and A. Stanghellini, *Nuovo Cimento 5*, 1305 (1957).

terms of the Chew-Low effective range formula, as well as the Serber form (3.13). They find:

$$\left.\begin{aligned} f^2 &= 0.095 \pm 0.006 \\ \omega_0^* &= 2.22 \pm 0.20 \end{aligned}\right\} \text{Chew-Low plot},$$

and (3.17)

$$\left.\begin{aligned} f^2 &= 0.107 \pm 0.007 \\ \omega_0^* &= 2.08 \pm 0.20 \end{aligned}\right\} \text{Serber plot}.$$

This gives a coupling constant somewhat larger than found above, but in agreement within the limits of error.

We remark here that our results (3.13) agree with the field-theoretical treatment for the 3,3 state, but not for the others—the values of r_α are not quantitatively correct, although their signs are. The proper field theory treatment shows that in general $r_1 = -r_3 - \frac{1}{4}r_2$, and under reasonable assumptions, $r_1 \simeq r_2 \simeq -0.8r_3$. The phase shifts for p-states other than the 3,3 state are very poorly determined by the experimental data, and no meaningful comparison with the effective range formula (3.13) can as yet be made.

3.2. Dispersion Relations

Another approach used to analyze the scattering data is by dispersion relations. We will see that this leads to a determination of the coupling constant f^2 in a manner similar to the effective range plot, although essentially independent of any theoretical model.

Dispersion relations, relations between the real part of a scattering amplitude and an integral over the imaginary part as a function of frequency, date back thirty years. Kronig and Kramers[24] first pointed out that for the coherent forward scattering of light, the real and imaginary parts of the scattering amplitude $f(\omega)$ were related by:

$$\mathrm{Re}[f(\omega) - f(0)] = \frac{2\omega^2}{\pi} P \int_0^\infty \frac{\mathrm{Im}\, f(\omega')}{\omega'(\omega'^2 - \omega^2)}\, d\omega', \tag{3.18}$$

where P means principal value. Kramers emphasized that the relation followed from the requirement of causality[25] alone, although the derivation was based on a particular model for the scattering of light. In recent

[24] R. Kronig, *J. Opt. Soc. Am. 12,* 546 (1926); H. A. Kramers, *Atti Congr. Intern. fis. Como 2,* 545 (1927).

[25] Loosely speaking, the cause always precedes the effect. All signals or interactions propagate with velocities not greater than the velocity of light. The requirement of causality thus means that a happening at one spacetime point can influence what happens at another point only if the separation of the two points is "time like," and further only if the first point lies in the backward light cone. Events at points with "space like" separations can not influence each other in any way.

years there has been a revival of interest within the context of high-energy physics.[26]

We will give a simple argument to show how the causality requirement gives rise to a dispersion relation of the type (3.18). Consider the forward scattering amplitude $f(\omega)$ for scattering of a real wave by some interaction. $f(\omega)$ can be thought of as the Fourier transform of a time-dependent amplitude $g(t)$:

$$f(\omega) = \int_{-\infty}^{\infty} g(t) e^{i\omega t}\, dt.$$

The causality requirement can be imposed by demanding $g(t) = 0$ for all $t < 0$, where we imagine a sharp wave packet reaching the interaction at $t = 0$. For a real wave packet, $g(t)$ will also be real. With this condition on $g(t)$ the Fourier integral becomes:

$$f(\omega) = \int_{0}^{\infty} g(t) e^{i\omega t}\, dt. \tag{3.19}$$

This is sufficient to determine the dispersion relations. $f(\omega)$ is divided into real and imaginary parts:

$$d(\omega) = \operatorname{Re} f(\omega) = \int_{0}^{\infty} g(t) \cos \omega t\, dt$$

and (3.20)

$$a(\omega) = \operatorname{Im} f(\omega) = \int_{0}^{\infty} g(t) \sin \omega t\, dt.$$

Since $g(t)$ is real, it is clear that $d(-\omega) = d(\omega)$, $a(-\omega) = -a(\omega)$. It is convenient to introduce the integral relation (for $t > 0$):

$$\cos \omega t = \frac{1}{\pi} P \int_{-\infty}^{\infty} d\omega' \frac{\sin \omega' t}{\omega' - \omega},$$

and substitute it into the right-hand side of the equation for $d(\omega)$:

$$d(\omega) = \int_{0}^{\infty} dt\, g(t) \frac{P}{\pi} \int_{-\infty}^{\infty} d\omega' \frac{\sin \omega' t}{\omega' - \omega}.$$

If the orders of integration are interchanged (allowable if certain restrictions are placed on $f(z)$), and the definition of $a(\omega)$ is employed, then the result is:

$$d(\omega) = \frac{1}{\pi} P \int_{-\infty}^{\infty} \frac{a(\omega')}{\omega' - \omega} d\omega'. \tag{3.21}$$

Finally the integral over negative frequencies is converted into one over

[26] See, for example, Gell-Mann, Goldberger, Thirring, *Phys. Rev.* *95*, 1612 (1954); M. L. Goldberger, *Phys. Rev.* *99*, 979 (1955); A. Salam, *Nuovo Cimento* *3*, 428 (1956); K. Symanzik, *Phys. Rev.* *105*, 743 (1957); Chew, Goldberger, Low, and Nambu, *Phys. Rev.* *106*, 1337, 1345 (1957).

positive frequencies using the odd character of $a(\omega)$. The end result is a "dispersion relation":

$$d(\omega) = \frac{2}{\pi} P \int_0^\infty \frac{\omega' \, a(\omega')}{\omega'^2 - \omega^2} \, d\omega'. \tag{3.22}$$

With slight further manipulation this will give the Kramers-Kronig relation (3.18) for forward scattering of light. One might wonder at the preference of the Kramers-Kronig form over (3.22). As a practical matter the point is that for large ω', the behavior of $a(\omega')$ may be such that the integral in (3.22) is not defined, whereas the integral in (3.18) converges. Such difficulty can be traced to the presence of delta function singularities in $g(t)$ at $t = 0$. Without detailed analysis one cannot determine such things in advance. A working procedure is to write down (3.22), and then perform the number of subtractions dictated by the asymptotic form of $a(\omega')$ for the problem at hand.

The quantity $a(\omega)$ is directly related to the total cross section through the "optical theorem":[27]

$$\sigma(\omega) = \frac{4\pi}{k} \, a(\omega), \tag{3.23}$$

where k is the wave number associated with the frequency ω. Consequently, in the dispersion relation the real part of the forward-scattering amplitude is determined by an integral over the total cross section.

The above discussion is obviously just an indication and not a proper proof of the dispersion relation. The quantum-mechanical proofs begin with the microscopic causality requirement that boson (fermion) field operators commute (anticommute) at space-like separations, a condition which is intuitively akin to our macroscopic ideas of causality.

3.3. Dispersion Relations for Pion-Nucleon Scattering

The application of dispersion relations to pion-nucleon forward scattering is complicated by (1) the charges of the pions and nucleons, (2) the non-vanishing mass of the pion. The first problem can be handled relatively straightforwardly.[28] If the $\pi^+ + p$ and $\pi^- + p$ scatterings are described by forward-scattering amplitudes:

$$F_\pm(\omega) = D_\pm(\omega) + iA_\pm(\omega), \tag{3.24}$$

and the following linear combinations are formed:

$$\begin{aligned} F_E(\omega) &= \tfrac{1}{2}(F_-(\omega) + F_+(\omega)) \\ F_O(\omega) &= \tfrac{1}{2}(F_-(\omega) - F_+(\omega)), \end{aligned} \tag{3.25}$$

[27] The optical theorem follows as a direct consequence of the unitarity of the scattering matrix. See B. A. Lippmann and J. Schwinger, *Phys. Rev.* *79*, 469 (1950), Eq. (1.75).

[28] See Goldberger, Miyazawa, and Oehme, *Phys. Rev.* *99*, 986 (1955).

then under the transformation $\omega \to -\omega$, the real and imaginary parts of the $\pi^+ + p$ and $\pi^- + p$ amplitudes transform into each other in such a way that:

$$F_E(-\omega) = F_E^*(\omega)$$

and (3.26)

$$F_O(-\omega) = -F_O^*(\omega).$$

This allows us to write down the two dispersion relations:

$$D_E(\omega) - D_E(\mu) = \frac{2k^2}{\pi} P \int_0^\infty \frac{\omega' A_E(\omega')}{k'^2(\omega'^2 - \omega^2)} d\omega' \tag{3.27}$$

and

$$D_O(\omega) = \frac{2\omega}{\pi} P \int_0^\infty \frac{A_O(\omega')}{\omega'^2 - \omega^2} d\omega', \tag{3.28}$$

where we have used a subtraction for $D_E(\omega)$ to assure convergence of the integral, and explicitly introduced the meson rest mass, $\omega = \mu = 1$.

Before these relations can be applied we must deal with the problem of the finite pion mass. The integrals in (3.27) and (3.28) go from $0 < \omega < \infty$, whereas the physical range is $\mu < \omega < \infty$. The range from $0 < \omega < \mu$ must be dealt with separately before the optical theorem can be used to get the remaining integrand from experiment. It is easy to show that $A(\omega)$ vanishes from $-\mu < \omega < \mu$, except for possible discrete points on that interval. The reason is that $A(\omega)$ involves only those processes which conserve energy (see the optical theorem). In the physical region scattering, either elastic or inelastic, can always occur, and $A(\omega)$ is different from zero. But in the unphysical region no energy-conserving process can occur in general. At certain discrete energies, however, a hypothetical process may be allowed by conservation of energy. In the scattering of a pion by a nucleon the absorption of the pion can occur with conservation of energy, provided the total energy of the pion and the nucleon in the center of mass system is equal to the nucleon mass M. This energy clearly lies in the unphysical region and is the only energy value in the range $0 < \omega < \mu$ at which there is a discrete contribution[29] to $A(\omega)$. The center of mass energy is given by

$$E_{CM} = \omega_q + \sqrt{q^2 + M^2},$$

so that the discrete contribution to the dispersion integrals comes at a pion energy ω_B such that

$$\omega_q + \sqrt{q^2 + M^2} = M.$$

This gives

$$\omega_B = \frac{1}{2M}. \tag{3.29}$$

[29] There is a corresponding discrete contribution on the interval $-\mu < \omega < 0$, but by using (3.26) we need only consider positive values of ω.

The imaginary part of the forward scattering amplitude for $\omega > 0$ can now be written $A(\omega) = C_B\,\delta(\omega - \omega_B) +$ continuum contribution for $\omega > \mu$, where the constant C_B can only be found from some theory. If we use the relativistic γ_5 theory (1.3) we find:

$$C_B = \pm\pi f^2, \tag{3.30}$$

where the plus and minus signs correspond to A_E and A_O, respectively, and $f^2 = \frac{1}{4\pi}\left(\frac{G}{2M}\right)^2$ is the *renormalized* unrationalized pseudovector coupling constant. This is essentially a *definition* of the coupling constant. It agrees with the definition of the static theory.

With the expression (3.30) for C_B and the optical theorem (3.23) we obtain the dispersion relations:

$$D_E(k) - D_E(0) = \frac{f^2}{M}\frac{k^2}{\omega^2 - \omega_B^2} + \frac{k^2}{2\pi^2} P\int_0^\infty \frac{dk'}{k'^2 - k^2}\left(\frac{\sigma_-(k') + \sigma_+(k')}{2}\right)$$

and (3.31)

$$D_O(k) = 2f^2\frac{\omega}{\omega^2 - \omega_B^2} + \frac{\omega}{2\pi^2} P\int_0^\infty \frac{k'^2\,dk'}{\omega'(k'^2 - k^2)}\left(\frac{\sigma_-(k') - \sigma_+(k')}{2}\right), \tag{3.32}$$

where the amplitudes and cross sections are in terms of the momentum k.

With the dispersion equations (3.31) and (3.32), or ones derived from them, numerous plots of experimental quantities can be made to determine various parameters. Anderson, Davidon, and Kruse[30] took linear combinations to get dispersion relations for the forward-scattering amplitudes for $T = \frac{3}{2}$ and $T = \frac{1}{2}$. They chose a coupling constant of $f^2 = 0.08$ to determine the leading terms in the dispersion relations, and evaluated the principal value integrals numerically with the observed total cross sections. These results, corresponding to the right-hand sides of (3.31) and (3.32), were then compared with the real parts of the forward-scattering amplitudes deduced from the phase shift analysis (see Section 2.3) to see whether the dispersion relations were satisfied. Figs. 4a and 4b show the type of results obtained for the $T = \frac{3}{2}$ and $T = \frac{1}{2}$ isotopic spin states. For convenience $2k_b^2 D(k)/k$ is plotted rather than $D(k)$, where k, k_b are the laboratory and center of momentum wave numbers respectively. The values of $D_\pm(0)$ needed in (3.31) were obtained from the s-wave phase shifts of (2.21). Anderson, Davidon, and Kruse were able to show that the dispersion relations were consistent with the Fermi or Yang sets of phase shifts, but not with any of the other proposed solutions. These two sets show a resonant behavior causing the abrupt change in sign of $D_3(k)$ at $\sim$180 Mev.

[30] Anderson, Davidon, and Kruse, *Phys. Rev.* *100,* 339 (1955).

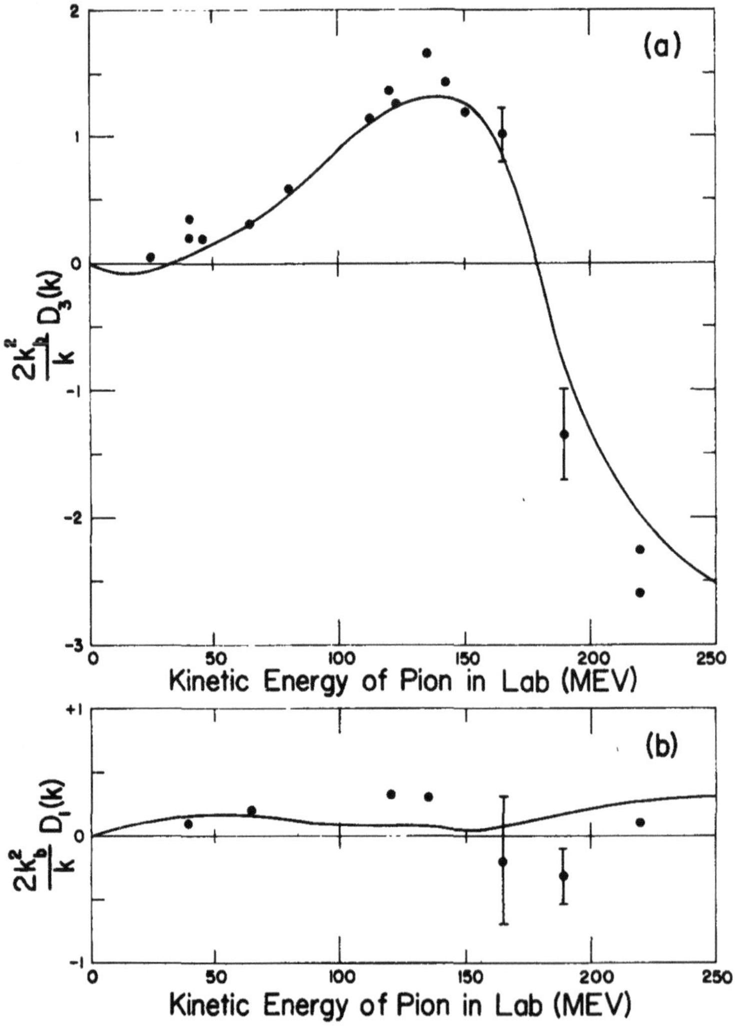

Fig. 4. Comparison of pion-nucleon scattering data by dispersion relations. The solid curve is obtained from the dispersion integral over the total cross sections with $f^2 = 0.08$. The points represent the values of $D(k)$ obtained from the phase shift analysis at various energies. The errors shown on two of the points are typical of the uncertainties on all the points. Fig. 4a is appropriate to the isotopic spin $T = \frac{3}{2}$ state, while Fig. 4b is for the $T = \frac{1}{2}$ state. (Based on Figs. 4 and 5 of Anderson, Davidon, and Kruse, footnote 30.)

Puppi and Stanghellini[23] have analyzed the latest data in essentially the same manner as Anderson, Davidon, and Kruse, but include data up to 400 Mev. For $D_3 = D_+$ they find essentially the same result as is shown in Fig. 4a, with the curve reaching a minimum at ~330 Mev and rising towards the axis at higher energies, and obtain agreement with experiment, provided $f^2 = 0.09 \pm 0.01$. From a similar plot for D_-, they find that the coupling constant should be of order one half of that for D_+ in order to obtain a fit. This argues for either (a) a breakdown of causality, or (b) lack of charge independence. The result is not conclusive, however, because of the large experimental errors for the small quantities in D_- and also the uncertainty in evaluating the principal value integrals involved.

On the basis of dispersion equations for the spin-flip amplitude (essentially $C(\omega)$ in (2.13)), Davidon and Goldberger[31] were able to exclude the Yang set of phase shifts because of erratic energy dependence. Thus the Fermi set of phase shifts are left as the only phase shifts satisfying the causality requirement.

The coupling constant can be evaluated accurately via the dispersion relations in various ways. One way is by use of (3.32). Using the identity

$$\frac{1}{\omega'^2 - \omega^2} = \frac{1}{\omega'^2} + \frac{\omega^2}{\omega'^2(\omega'^2 - \omega^2)},$$

(3.32) can be rewritten in the form:

$$D_O(\omega) - \frac{\omega^3}{2\pi^2} P \int_1^\infty d\omega' \frac{k'\frac{1}{2}(\sigma_- - \sigma_+)}{\omega'^2(\omega'^2 - \omega^2)}$$
$$= 2f^2 \frac{\omega}{\omega^2 - \omega_B^2} + \frac{\omega}{2\pi^2} P \int_1^\infty d\omega' \frac{k'}{\omega'^2} \left(\frac{\sigma_- - \sigma_+}{2}\right).$$

Multiplication across by $\omega/(\omega^2 - \omega_B^2)$ yields the result:

$$\frac{\omega^2 - \omega_B^2}{\omega} \left[D_O(\omega) - \frac{\omega^3}{2\pi^2} P \int_1^\infty d\omega' \frac{k'\frac{1}{2}(\sigma_- - \sigma_+)}{\omega'^2(\omega'^2 - \omega^2)} \right] \tag{3.33}$$
$$= 2f^2 + \frac{\omega^2 - \omega_B^2}{2\pi^2} P \int_1^\infty d\omega' \frac{k'}{\omega'^2} \left(\frac{\sigma_- - \sigma_+}{2}\right).$$

A plot of the left-hand side of (3.33) from experiment as a function of ω^2 should be a straight line with intercept $2f^2$ at $\omega^2 = 0$. The advantage of (3.33) stems from the fact that the integral on the left-hand side converges very rapidly at high energies and so is insensitive to the exact asymptotic behavior of the cross sections. Haber-Schaim[32] finds in this way $f^2 = 0.08 \pm 0.01$, in good agreement with the effective range method of Section 3.1.

Gilbert* has derived a different set of dispersion relations which are more or less the inverse of (3.31) and (3.32) in that they relate the imaginary part of the scattering amplitude (actually $A(\omega)/k$) to an integral over

* W. Gilbert, *Phys. Rev.* *108*, 1078 (1957).

[31] W. C. Davidon and M. L. Goldberger, *Phys. Rev.* *104*, 1119 (1956).

[32] U. Haber-Schaim, *Phys. Rev.* *104*, 1113 (1956).

the real part. Using his relations for the spin-flip amplitude, Gilbert is able to make an independent evaluation of the coupling constant, obtaining $f^2 = 0.084$, in good agreement with Haber-Schaim. Further application of his relations allows him to determine the low-energy behavior of the *s*-wave phase shifts in terms of dispersion integrals involving the dominant *p*-wave interaction. With a coupling constant of ~0.083 the calculated *s*-wave phase shifts are in agreement with the experimental values (2.21).

Dispersion relations form a very powerful method of analysis and have been applied widely to pion-nucleon scattering in the forward direction, as well as to Compton scattering by protons.[33] Various proofs have been given of the relations for both forward scattering and non-forward scattering.[34] The non-forward dispersion relations (except for the special case of the spin-flip term) have not been used because of the difficulty of handling the "unphysical" region in the integrals (this region now depends on the momentum transfer in a complicated way). Oehme[35] has derived the Low equations for *s*-waves as well as *p*-waves for the static model from the relativistic dispersion relations by assuming that partial waves greater than $l = 1$ could be neglected (see also Chew, Goldberger, Low, Nambu, footnote 26). Recently Goldberger, Nambu, and Oehme[36] have derived dispersion relations for nucleon-nucleon scattering, and hope to use them as a method of obtaining the nucleon-nucleon interaction. Khuri[37] and others have obtained dispersion relations for nonrelativistic scattering problems where the causality requirement is replaced by conditions on the form of the scattering potential.

---

[33] Gell-Mann, Goldberger, and Thirring, footnote 26; R. H. Capps, *Phys. Rev. 106*, 1031 (1957).

[34] K. Symanzik, footnote 26; N. N. Bogoliubov, *Report to the International Physics Conference at Seattle*, 1956; Bremermann, Oehme, and Taylor, *Phys. Rev. 109*, 2178 (1958).

[35] R. Oehme, *Phys. Rev. 102*, 1174 (1956).

[36] Goldberger, Nambu, and Oehme, *Annals of Physics 2*, 226 (1957).

[37] N. N. Khuri, *Phys. Rev. 107*, 1148 (1957).

CHAPTER 4

Photoproduction of Pions from Nucleons

The production of pions by gamma radiation incident on nucleons has been studied extensively in the energy range from threshold to 450 Mev, and less thoroughly up to 1 Bev. The best known processes are

$$\gamma + p \rightarrow p + \pi^0$$
$$\gamma + p \rightarrow n + \pi^+ ,$$

although one can get at the reaction $\gamma + n \rightarrow p + \pi^-$ by studying photoproduction in deuterium. The observed total cross sections for

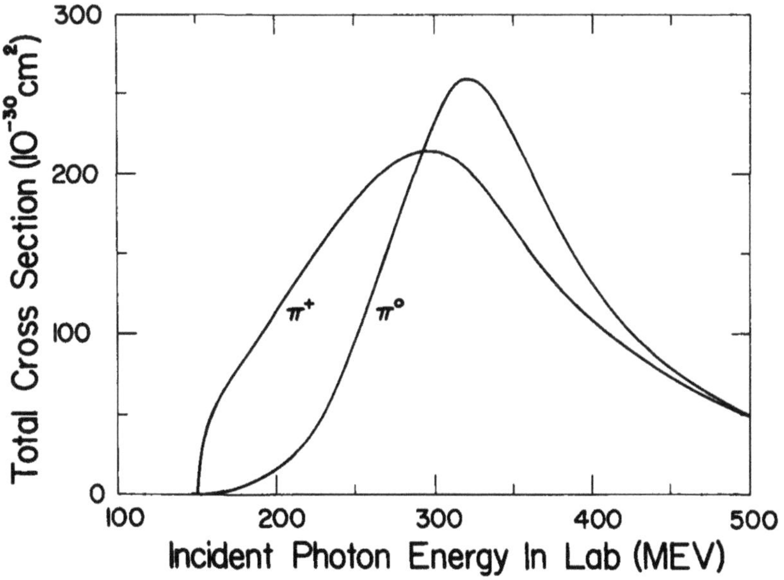

Fig. 5. Total cross sections for the photoproduction of neutral and charged pions from hydrogen as a function of the incident photon energy.

photoproduction of neutral and charged pions from hydrogen are shown in Fig. 5. The cross sections exhibit a resonance behavior like the scattering in the photon energy range 300–350 Mev, corresponding to a pion kinetic energy of ~180 Mev. We therefore expect the (3,3) resonance to play an important role. Near threshold, the neutral pion cross section rises very slowly (roughly proportional to $(E_\gamma - E_0)^{3/2}$), while the positive pion cross section rises more rapidly, like $(E_\gamma - E_0)^{1/2}$. This indicates that for charged pions s-wave production is appreciable near threshold, while for neutral pions p-wave production dominates at all energies (at least below 400 Mev). Recent experiments at the California Institute of Technology and Cornell University have brought to light a higher energy resonance (or perhaps more than one) in the positive pion photoproduction at about 700 Mev photon energy. The neutral pion photoproduction shows a much smaller peak. The resonance is smaller in height and wider than the (3,3) resonance shown in Fig. 5. It is probably in the $T = \frac{1}{2}$ state and may involve higher partial waves than $l = 1$. We will restrict our discussion to the neighborhood of the (3,3) resonance and below.

One way of discussing the problem is by means of a multipole expansion.[38] Thus one would consider the first few multipoles, say $E1$, $M1$, and $E2$. The $E1$ absorption would lead to s-wave pions in the final state (i.e. the $s\frac{1}{2}$ scattering state), while $M1$ and $E2$ absorption would give p-wave pions (the $p_{1/2}$ and $p_{3/2}$ states). When the possible states of $T = \frac{1}{2}, \frac{3}{2}$ are considered, there are 8 matrix elements involved (2 each for $E1$ and $E2$, 4 for $M1$). We will not use the multipole approach, but will treat the problem in a manner similar to that used in Section 3.1 on scattering. This has the advantage that we can draw from our experience there. In addition, one of the amplitudes which is equivalent to many multipoles in the charged pion photoproduction can be handled reasonably accurately in closed form. Only the static limit will be treated. Effects such as nucleon recoil (which contributes to the difference between the cross sections for photoproduction of positive and negative pions from deuterium) will be ignored.

4.1. Equations of Motion in the Presence of the Electromagnetic Field

In the discussion so far, the electromagnetic field has been absent and the pion field satisfied (1.12), while the nucleon variables $\vec{\sigma}$ and $\boldsymbol{\tau}$ satisfied (1.14) and (1.15). In order to incorporate the electromagnetic field $A_\nu = (\vec{A}, i\Phi)$, the pion field must be separated into its neutral and charged parts:

$$\phi_0 = \phi_3, \qquad \phi_\pm = \mp \frac{1}{\sqrt{2}}(\phi_1 \pm i\phi_2). \tag{4.1}$$

The neutral pion field is not affected directly by A_ν and so satisfies (1.12):

$$(\Box^2 - 1)\phi_0 = -F\tau_3\vec{\sigma}\cdot\vec{\nabla}\rho. \tag{4.2}$$

[38] B. T. Feld, *Phys. Rev. 89*, 330 (1953); Watson, Keck, Tollestrup, and Walker, *Phys. Rev. 101*, 1159 (1956).

The charged pion field is, however, directly coupled to the electromagnetic field. This coupling is taken into account by the replacement in the Lagrangian density:

$$\frac{\partial \phi_\pm}{\partial x_\nu} \rightarrow \left(\frac{\partial}{\partial x_\nu} \mp ieA_\nu\right)\phi_\pm. \tag{4.3}$$

This is equivalent to the well known classical substitution $p_\nu \rightarrow \left(p_\nu - \frac{e}{c} A_\nu\right)$ for a particle with charge e. When this substitution is made in the free and interaction Lagrangian densities (see (1.9) and (1.11)) the result can be expressed as an additional electromagnetic interaction Lagrangian density:

$$\mathscr{L}_{e-m} = -ie \sum_{\nu=1}^{4} \left(\phi_+ \frac{\partial \phi_-}{\partial x_\nu} - \phi_- \frac{\partial \phi_+}{\partial x_\nu}\right) A_\nu + e^2 A^2 \phi_+ \phi_-$$
$$- ieF\sqrt{2}(\tau_+ \phi_- + \tau_- \phi_+)(\vec{\sigma} \cdot \vec{A})\rho(\vec{r}), \tag{4.4}$$

where $\tau_\pm = \frac{1}{2}(\tau_1 \pm i\tau_2)$ are operators which turn neutron into proton and vice versa. The first term is of the form of a current density times the vector potential, namely the pion current density. It represents the direct coupling of the charged pion field to the electromagnetic field. The second term is the "double photon" term, and does not enter into processes involving emission or absorption of single quanta. The third term arises from the gradient coupling in the interaction Lagrangian as a consequence of the gauge invariant replacement (4.3). It gives rise to s-wave pion production, and is important near threshold for charged pions.[39]

If the electromagnetic interaction Lagrangian (4.4) is combined with the other terms, and the equations of motion for the charged fields obtained according to (1.10), the result is:

$$(\square^2 - 1)\phi_\pm = \pm\sqrt{2}F\tau_\pm \vec{\sigma} \cdot \vec{\nabla} \rho \pm 2ie\vec{A} \cdot \vec{\nabla}\phi_\pm - \sqrt{2}ieF\tau_\pm(\vec{\sigma} \cdot \vec{A})\rho, \tag{4.5}$$

where it has been assumed that the electromagnetic field A_ν is actually a radiation field described by a vector potential $\vec{A}$. In discussing photoproduction by means of (4.5) we must think of it as an operator equation in the nucleon variables, and eventually take matrix elements of it between the appropriate nucleon states. From the properties of $\tau_\pm$ it is clear that in order to have charge conservation in the emission of pions

[39] Because of the extended source density $\rho(r)$, the third term in (4.4) is not quite gauge invariant as it stands, but must have line currents added (see R. H. Capps and W. G. Holladay, *Phys. Rev. 99*, 931 (1955)). For a compact source distribution, however, the added contribution is not important.

(described by $\phi_{\pm}$) we must form matrix elements with the initial nucleon state on the left and the final nucleon state on the right.

Equation (4.2) for the neutral pions and (4.5) for the charged pions can be combined into one isotopic spin vector equation:

$$(\Box^2 - 1)\boldsymbol{\phi} = -F\boldsymbol{\tau}\vec{\sigma}\cdot\vec{\nabla}\rho + 2e\vec{A}\cdot\vec{\nabla}(\boldsymbol{\epsilon}_3 \times \boldsymbol{\phi}) + eF(\boldsymbol{\epsilon}_3 \times \boldsymbol{\tau})(\vec{\sigma}\cdot\vec{A})\rho, \tag{4.6}$$

where $\boldsymbol{\epsilon}_3$ is a unit vector in the 3-direction in isotopic spin space. The photoproduction of pions of different charge states is obtained by taking the isotopic spin scalar product of (4.6) with various combinations of the three unit vectors, $\boldsymbol{\epsilon}_1$, $\boldsymbol{\epsilon}_2$, $\boldsymbol{\epsilon}_3$: for neutral pions, $\boldsymbol{\epsilon}_3$; for positive pions, $-\frac{1}{\sqrt{2}}(\boldsymbol{\epsilon}_1 + i\boldsymbol{\epsilon}_2)$; for negative pions, $\frac{1}{\sqrt{2}}(\boldsymbol{\epsilon}_1 - i\boldsymbol{\epsilon}_2)$; in complete correspondence with ordinary angular momentum.

The contribution of each of the three terms on the right-hand side of (4.6) to the photoproduction process will be considered separately.

4.2. Nucleon Magnetic Moment Contributions and Neutral Pion Photoproduction

At first glance it would appear that the first term in (4.6) does not contribute to the photoproduction, since it does not involve the electromagnetic field. It is, in fact, exactly the same term as we had in the scattering problem, (1.12) or (2.1). The difference lies in the fact that the time dependence of $\vec{\sigma}$ and $\boldsymbol{\tau}$ is different in the presence of the electromagnetic field. One can think of the incident photon interacting with the nucleon's magnetic moment and causing a precession which gives rise to emission of mesons.

Just as in going from (3.1) to (3.5) in the scattering problem, we now consider approximate solutions of the equations of motion for $\vec{\sigma}$ and $\boldsymbol{\tau}$ in the presence of the electromagnetic field as well as the pion field. In our linear approximation the effects of the pion field and the electromagnetic field on $\vec{\sigma}$ and $\boldsymbol{\tau}$ can be added together. Thus there will be the time dependent terms $\vec{\sigma}_1$ and $\boldsymbol{\tau}_1$, as before (see (3.2)), as well as additional terms from the interaction between the nucleon's magnetic moment and the magnetic field:

$$H = -\vec{\mu}\cdot\vec{B}. \tag{4.7}$$

This is well defined only for static fields and moments, but for moderate frequencies it is a sufficiently good approximation to continue to use the static moments for $\vec{\mu}$. Thus it will be assumed that for time-varying fields:

$$H = -\left[\frac{\mu_p + \mu_n}{2} + \frac{\mu_p - \mu_n}{2}\tau_3\right]\frac{e}{2M}\vec{\sigma}\cdot\vec{\nabla}\times\vec{A}, \tag{4.8}$$

where μ_p and μ_n are the static proton and neutron magnetic moments in units of $e/2M$ ($\mu_p = 2.79$, $\mu_n = -1.91$). The interaction (4.8) is, of course, an $M1$ interaction and so will give rise to p-wave mesons. It is useful to divide it into two parts, an isotopic spin scalar and an isotopic spin vector:

$$H_s = -\left(\frac{\mu_p + \mu_n}{2}\right) \frac{e}{2M} \vec{\sigma} \cdot \vec{\nabla} \times \vec{A} \tag{4.9}$$

and

$$H_v = -\left(\frac{\mu_p - \mu_n}{2}\right) \tau_3 \frac{e}{2M} \vec{\sigma} \cdot \vec{\nabla} \times \vec{A}. \tag{4.10}$$

The reason for this division is that the scalar term cannot, in the photoproduction process, give rise to pions in the final state $T = \frac{3}{2}$, whereas the vector term can reach this state. In the scattering problem we saw that the $T = \frac{3}{2}$ p-state was exceptional, and the correct behavior of the scattering in that state was far from the Born approximation, while for other states the Born approximation was qualitatively correct. So also here we expect that it will be sufficient to treat H_s in lowest order approximation, while for H_v more careful consideration will be necessary. Fortunately, as will be seen almost immediately, H_v is closely related to the scattering and can be expressed in terms of its known properties.

The contribution of H_s to the photoproduction amplitude will be considered first. If the electromagnetic field is normalized to unit incident flux, then

$$\vec{A} = \vec{\varepsilon} \sqrt{\frac{2\pi}{\omega}}\, e^{i(\vec{k}\cdot\vec{r} - \omega t)}, \tag{4.11}$$

where $\vec{\varepsilon}$ is the usual polarization vector, $\vec{k}$ is the photon wave vector, and $\omega = k$ is its frequency. By using $\dot{\vec{\sigma}} = \frac{1}{i}[\vec{\sigma}, H]$ as before, it is easy to obtain the first order approximation:

$$(\nabla^2 + q^2)\boldsymbol{\phi} = \frac{4\pi e f}{\sqrt{2\omega}} \left(\frac{\mu_p + \mu_n}{2M}\right) \boldsymbol{\tau} \left[\frac{\vec{\sigma} \times (\vec{k} \times \vec{\varepsilon})}{\omega}\right] \cdot \vec{\nabla}\rho \tag{4.12}$$

as the equation for the photopion field due to H_s. $\boldsymbol{\phi}$ is a "scattered" wave, with the boundary condition:

$$\boldsymbol{\phi} \to f(\theta, \varphi) \frac{e^{iqr}}{r}.$$

Using the usual Green's function technique, (4.12) can be solved to yield:

$$f_s = \boldsymbol{\tau} e f \left(\frac{\mu_p + \mu_n}{2M}\right) \frac{v(\vec{q})}{\sqrt{2\omega}} \frac{i\vec{\sigma} \cdot [\vec{q} \times (\vec{k} \times \vec{\varepsilon})]}{\omega}. \tag{4.13}$$

For low-energy pions it is sufficient to put $v(\vec{q}) \simeq 1$.

The radial flux (number of pions per unit solid angle per second) is given by $2q|(f)|^2$. With unit incident photon flux the differential cross section is

$$\frac{d\sigma}{d\Omega} = 2q|f|^2. \tag{4.14}$$

For neutral pion production on protons the contribution of H_s to the amplitude is:

$$f_s^0 = \frac{ef}{\sqrt{2\omega}}\left(\frac{\mu_p + \mu_n}{2M}\right)\frac{i\vec{\sigma}\cdot[\vec{q}\times(\vec{k}\times\vec{\varepsilon})]}{\omega}, \tag{4.15}$$

while for positive or negative pions (from protons or neutrons) it is:

$$f_s^{\pm} = \mp\sqrt{2}f_s^0. \tag{4.16}$$

It will turn out that this p-wave contribution due to H_s is negligible compared to that from H_v.

We now consider the much more important magnetic moment term H_v. By using the equation of motion for $\vec{\sigma}$ and $\boldsymbol{\tau}$ and looking for terms with time dependence exp $(-i\omega t)$, just as in the scattering problem, one finds that the contribution of H_v to the inhomogeneous term $(-F\boldsymbol{\tau}\vec{\sigma}\cdot\vec{\nabla}\rho)$ is:

$$\frac{4\pi ef}{\sqrt{2\omega}}\left(\frac{\mu_p - \mu_n}{2M}\right)\frac{1}{\omega}[\boldsymbol{\epsilon}_3\vec{\sigma}\times(\vec{k}\times\vec{\varepsilon}) + (\boldsymbol{\tau}\times\boldsymbol{\epsilon}_3)\vec{k}\times\vec{\varepsilon}]\cdot\vec{\nabla}\rho\,.$$

If we define the momentum vector $\vec{q}_0$ to have the direction $(\vec{k}\times\vec{\varepsilon})$, but the magnitude $q = \sqrt{k^2 - 1}$ of the emitted pion, this result can be written:

$$-\frac{8\pi f^2}{\omega}A[\boldsymbol{\epsilon}_3\vec{\sigma}\times\vec{q}_0 + (\boldsymbol{\tau}\times\boldsymbol{\epsilon}_3)\vec{q}_0]\cdot\vec{\nabla}\rho,$$

where

$$A = -\frac{e}{f}\left(\frac{\mu_p - \mu_n}{4M}\right)\frac{1}{\sqrt{2\omega}}\frac{k}{q}\,. \tag{4.17}$$

When this electromagnetic interaction term is combined with the pion interaction term on the right-hand side of (3.5), we obtain the somewhat involved equation for the photopion field due to H_v:

$$\begin{aligned}(\nabla^2 + q^2)\boldsymbol{\Phi}_v = &-\frac{8\pi f^2}{i\omega}A(\vec{\nabla}\rho)\cdot[\boldsymbol{\epsilon}_3(\vec{\sigma}\times i\vec{q}_0) + (\boldsymbol{\tau}\times\boldsymbol{\epsilon}_3)i\vec{q}_0] \\ &-\frac{8\pi f^2}{i\omega}(\vec{\nabla}\rho)\cdot\int[\vec{\sigma}\times\vec{\nabla}'\boldsymbol{\Phi}_v + \vec{\nabla}'(\boldsymbol{\tau}\times\boldsymbol{\Phi}_v)]\rho(\vec{r}')\,d\vec{r}'.\end{aligned} \tag{4.18}$$

There is, however, a marked similarity between the first and second terms on the right-hand side. In fact, if a plane wave is inserted in the second term for $\boldsymbol{\Phi}_v$, we get a result proportional to the first term. Remembering that $\boldsymbol{\phi}_v$, the photoproduced field, is a "scattered" wave in the sense that asymptotically it is a spherically diverging wave, we can define a hypothetical pion field:

$$\boldsymbol{\phi} = \boldsymbol{\epsilon}_3 \frac{A}{v(\vec{q}_0)} e^{i\vec{q}_0\cdot\vec{r}} + \boldsymbol{\phi}_v, \tag{4.19}$$

and see immediately that (4.18) for the photoproduced field is equivalent to the equation:

$$(\nabla^2 + q^2)\boldsymbol{\phi} = -\frac{8\pi f^2}{i\omega}(\vec{\nabla}\rho)\cdot\int[\vec{\sigma}\times\vec{\nabla}'\boldsymbol{\phi} + \vec{\nabla}'(\boldsymbol{\tau}\times\boldsymbol{\phi})]\rho(\vec{r}')\,d\vec{r}',$$

which is just (3.5) for the scattering problem. Thus photoproduction by the interaction H_v is entirely *equivalent* to the *scattering of a neutral pion* by a nucleon, with the modification that the incident wave is normalized with a coefficient $A/v(\vec{q}_0)$. The photoproduction of neutral pions from protons corresponds to the scattering $\pi^0 + p \rightarrow \pi^0 + p$, while the production of positive pions is equivalent to $\pi^0 + p \rightarrow \pi^+ + n$. This correspondence allows us to write down directly the photoproduction amplitude due to H_v in terms of the scattering amplitudes. If f_{scatt} is the scattering amplitude as defined in Section 2.3 for either the direct or exchange scattering of neutral pions on protons, it is clear from (4.19) that the photoproduction amplitude due to H_v is:

$$f_v = Af_{\text{scatt}}(\vec{q}_0, \vec{q}), \tag{4.20}$$

assuming that $v(\vec{q}_0) \simeq 1$, as usual. An alternative way to get this result is to compare matrix elements of H_v (see (4.10)) with the matrix elements V_q (see (1.19)).

For neutral pion photoproduction the contributions from H_s and H_v are all that occur, and hence the complete amplitude is given by the sum of (4.15) and (4.20). In the region of appreciable cross section the resonant enhancement of f_{scatt} means that (4.15) is numerically small compared to (4.20). Consequently, to an excellent approximation the neutral pion production cross section can be written as:

$$\frac{d\sigma}{d\Omega}(\vec{k}, \vec{\varepsilon}, \vec{q}) = \frac{e^2}{f^2}\left(\frac{\mu_p - \mu_n}{4M}\right)^2 \frac{k}{q}\left(\frac{d\sigma(\vec{q}_0, \vec{q})}{d\Omega}\right)_{\pi_0\rightarrow\pi_0}, \tag{4.21}$$

where $\vec{q}_0 = \frac{q}{k}(\vec{k}\times\vec{\varepsilon})$. For completeness and explicitness we record the actual amplitude f_v for neutral pion production in terms of the p-wave scattering parameters. (This can be obtained from (2.9), (2.13), and (2.18)):

$$f_v(\pi^0) = -\frac{e}{f}\left(\frac{\mu_p - \mu_n}{4M}\right)\frac{1}{\sqrt{2\omega}}\left[\vec{q}\cdot(\vec{k}\times\vec{\varepsilon})\left(\frac{a_1 + 4a_2 + 4a_3}{3q^2}\right) + i\vec{\sigma}\cdot(\vec{q}\times(\vec{k}\times\vec{\varepsilon}))\left(\frac{a_1 + a_2 - 2a_3}{3q^2}\right)\right], \tag{4.22}$$

where the a's are the scattering amplitudes (2.12), and the subscripts are the same as used in Section 3.1. For positive pions from protons the result has the same form, with the first linear combination of scattering amplitudes replaced by $-\sqrt{2}\,(a_1 + a_2 - 2a_3)$, and the second replaced by $-\sqrt{2}\,(a_1 - 2a_2 + a_3)$.

The photoproduction of neutral pions from hydrogen has been studied in a number of laboratories. The behavior of the total cross section as a function of photon energy was shown in Fig. 5. The resonant behavior at a photon energy of 320 Mev (equivalent to a pion kinetic energy in the

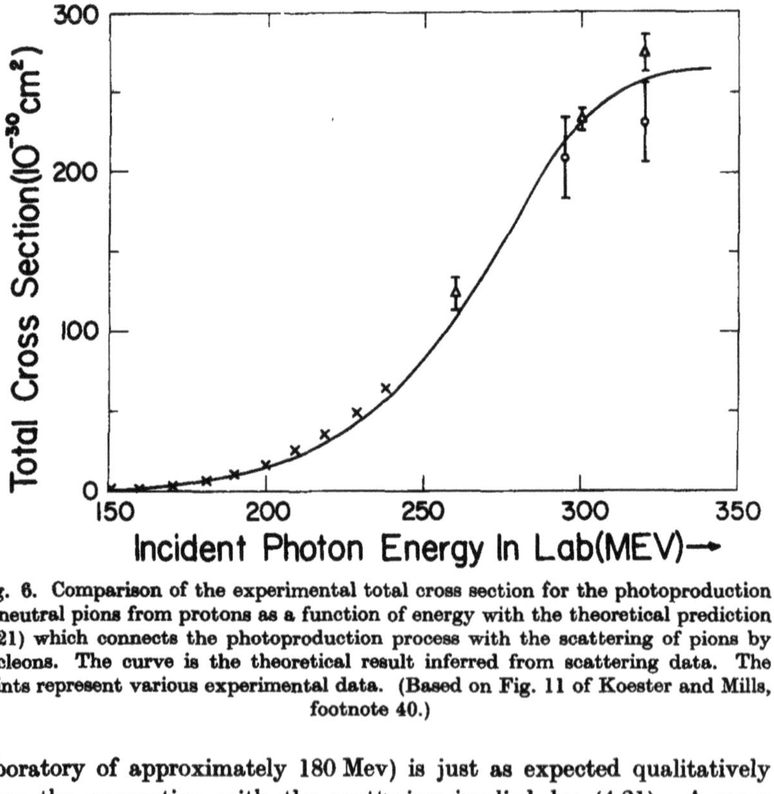

Fig. 6. Comparison of the experimental total cross section for the photoproduction of neutral pions from protons as a function of energy with the theoretical prediction (4.21) which connects the photoproduction process with the scattering of pions by nucleons. The curve is the theoretical result inferred from scattering data. The points represent various experimental data. (Based on Fig. 11 of Koester and Mills, footnote 40.)

laboratory of approximately 180 Mev) is just as expected qualitatively from the connection with the scattering implied by (4.21). A more quantitative comparison is shown in Fig. 6, where the total cross section predicted by (4.21) is compared with experimental data[40] up to 350 Mev. Considering the experimental uncertainties, the comparison is quite good.

Near the resonant energy it is clear that the 3,3 state plays the dominant role. If it is assumed in (4.22) that all amplitudes are negligible

[40] J. L. Koester, Jr. and F. E. Mills, *Phys. Rev.* *105*, 1900 (1957). See also McDonald, Peterson, and Corson, *Phys. Rev.* *107*, 577 (1957).

except a_3, the angular distribution for neutral pion photoproduction by unpolarized photons takes the form:

$$\frac{d\sigma}{d\Omega} \simeq \frac{2}{9}\frac{e^2}{f^2}\left(\frac{\mu_p - \mu_n}{4M}\right)^2 \frac{k \sin^2 \delta_{33}}{q^3}(2 + 3 \sin^2 \theta). \tag{4.23}$$

This is in rough accord with the data at photon energies from 300 to 350 Mev, but at lower energies the distribution has the unsymmetrical form $(A + B \cos \theta + C \cos^2 \theta)$, with B comparable to A and C.

4.3. Photoproduction of Charged Pions

In the photoproduction of charged pions all three terms in (4.6) contribute. The amplitude due to the nucleon magnetic moment has already been considered, with the results (4.16) and (4.22). We now turn to the second and third source terms in (4.6): the pion current contribution, and the s-wave "gauge" contribution. In accord with our general arguments that we should handle with precision only terms which give pions in the 3,3 state, we will treat the s-wave term only in lowest order approximation, even though it involves both $T = \frac{1}{2}$ and $T = \frac{3}{2}$ states. The pion current interaction will also be treated in lowest order, even though it involves p-wave pions, for reasons that will become more convincing after the elementary result is obtained.

In the pion current term in (4.6) it will be noted that both $\vec{A}$ and $\boldsymbol{\phi}$ appear. Consequently, care must be taken in handling the time dependences involved. As a first approximation it is assumed that $\vec{\sigma}$ and $\boldsymbol{\tau}$ are constant, while $\vec{A}$ is given by (4.11), and the pion field is written as a static field $\boldsymbol{\phi}_0$ plus a small time-dependent term $\boldsymbol{\phi}_1 e^{-i\omega t}$, which is the photoproduced field due to the second and third terms in (4.6). The equation of motion for $\boldsymbol{\phi}_0$ is just the static equation:

$$(\nabla^2 - 1)\boldsymbol{\phi}_0 = -F\boldsymbol{\tau}\vec{\sigma} \cdot \vec{\nabla}\rho, \tag{4.24}$$

while the time-dependent field satisfies the equation:

$$(\nabla^2 + q^2)\boldsymbol{\phi}_1 = e\sqrt{\frac{2\pi}{\omega}}\, e^{i\vec{k}\cdot\vec{r}}\,[F(\vec{\sigma} \cdot \vec{\varepsilon})(\boldsymbol{\epsilon}_3 \times \boldsymbol{\tau})\rho + 2\vec{\varepsilon} \cdot \vec{\nabla}(\boldsymbol{\epsilon}_3 \times \boldsymbol{\phi}_0)]. \tag{4.25}$$

The first term on the right gives the s-wave production; the second is the pion current contribution.

The amplitude for s-wave photoproduction is, in complete analogy to the handling of H_s in Section 4.2, found to be:

$$f_S = -\frac{ef}{\sqrt{2\omega}}(\vec{\sigma} \cdot \vec{\varepsilon})(\boldsymbol{\epsilon}_3 \times \boldsymbol{\tau})v(\vec{q} - \vec{k}). \tag{4.26}$$

This corresponds to an amplitude for positive pions from protons:

$$f_S^+ = \sqrt{2}\frac{ef}{\sqrt{2\omega}}(i\vec{\sigma}\cdot\vec{\varepsilon}), \tag{4.27}$$

where $v(\vec{q}-\vec{k})$ has been approximated by unity.

To find the pion current amplitude it is necessary to specify the static field $\boldsymbol{\phi}_0$. It is a good enough approximation to treat the nucleon as a point source. Then the static field is given by (1.16) of Section 1.3. When this static field is inserted into (4.25), one can calcuate the first order approximation to the pion current photoproduction amplitude:

$$f_\pi = -\frac{2ef}{\sqrt{2\omega}}\frac{(\vec{\varepsilon}\cdot\vec{q})\vec{\sigma}\cdot(\vec{k}-\vec{q})}{1+(\vec{k}-\vec{q})^2}(\boldsymbol{\epsilon}_3\times\boldsymbol{\tau}), \tag{4.28}$$

which gives the positive pion amplitude:

$$f_\pi^+ = 2\sqrt{2}\frac{ef}{\sqrt{2\omega}}\frac{i\vec{\sigma}\cdot(\vec{k}-\vec{q})(\vec{\varepsilon}\cdot\vec{q})}{1+(\vec{k}-\vec{q})^2}. \tag{4.29}$$

This is the lowest order result and one can well ask about corrections due to the 3,3 resonance. Chew and Low[3] have shown that these corrections are not vital. There is a resonant enhancement of f_π, but the enhancement of f_v (4.22) is much greater. Consequently, we lose little by being content with (4.29). Qualitatively, the explanation is that the photopions produced by the pion current term are "almost real" pions, out in the static cloud around the nucleon where they emerge with little secondary scattering, whereas the pions due to H_v are produced deep inside and undergo considerable multiple scattering before emerging.

The complete amplitude for photoproduction of positive pions from protons is the sum of the various terms, f_s, f_v, f_S, and f_π, given by (4.16), (4.22), (4.27), and (4.29):

$$\begin{aligned} f(\gamma\to\pi^+) \simeq \frac{\sqrt{2}ef}{\sqrt{2\omega}}\Bigg[& i\vec{\sigma}\cdot\vec{\varepsilon} + \frac{2i\vec{\sigma}\cdot(\vec{k}-\vec{q})(\vec{\varepsilon}\cdot\vec{q})}{1+(\vec{k}-\vec{q})^2} \\ & + \left(\frac{\mu_p-\mu_n}{4Mf^2}\right)\left\{\vec{q}\cdot(\vec{k}\times\vec{\varepsilon})\left(\frac{a_1+a_2-2a_3}{3q^2}\right)\right. \\ & \left. + i\vec{\sigma}\cdot(\vec{q}\times(\vec{k}\times\vec{\varepsilon}))\left(\frac{a_1-2a_2+a_3}{3q^2}\right)\right\} \\ & - \left(\frac{\mu_p+\mu_n}{4M}\right)\frac{i\vec{\sigma}\cdot(\vec{q}\times(\vec{k}\times\vec{\varepsilon}))}{\omega}\Bigg]. \end{aligned} \tag{4.30}$$

The total cross section for photoproduction of positive pions from protons is shown in Fig. 5. The rapid rise of the cross section above threshold is due to the *s*-wave contribution in (4.30). Near threshold the angular distribution is isotropic and the total cross section is given by:

$$\sigma(\gamma \rightarrow \pi^+) \simeq 8\pi e^2 f^2 \left(\frac{q}{\omega}\right). \tag{4.31}$$

This gives another method of evaluating the pion-nucleon coupling constant f^2. The value so obtained is in fair accord with the values from the scattering data. Near 300 Mev the 3,3 resonance makes itself felt, and the contribution of H_v is the dominant one in the amplitude (4.30), although the other terms are appreciable enough to give front-to-back asymmetries in the angular distributions. If the angles are not too near 0°, the angular distribution can be fitted with a distribution $(A + B\cos\theta + C\cos^2\theta)$, where $A > 0, C < 0$, and B is negative below ~340 Mev and positive above that energy.

For neutral pions the angular distributions in Section 4.2 were discussed in terms of powers of $\cos\theta$ up to $\cos^2\theta$. This implied that only *s*-wave and *p*-wave pions were produced, as is consistent with the connection between neutral pion photoproduction and scattering, and that only a small number of electromagnetic multipoles were involved. Initially, the angular distribution data on charged pion photoproduction was interpreted in this same way,[38] using powers of $\cos\theta$ up to $\cos^2\theta$. It is clear, however, from (4.29) that the pion current contribution cannot be expressed as a finite series in powers of $\cos\theta$, corresponding to the fact that it represents the sum of a large number of multipoles. Thus, the denominator is:

$$1 + (\vec{k} - \vec{q})^2 = 2\omega^2\left(1 - \frac{v}{c}\cos\theta\right), \tag{4.32}$$

where v is the velocity of the emitted pion. The pion current term will be most important at small angles where it will cause a departure from the simple three term polynominal in $\cos\theta$, especially for $v \sim c$.

Recently, observations have been made[41] at sufficiently small angles that the pion current contribution has been identified explicitly. Fig. 7 illustrates the type of result obtained for the angular distribution. The data of Fig. 7 are those of Malmberg and Robinson[42] at 225 Mev incident photon energy. The experimental points have been normalized to fit the calculated curve at large angles. The solid curve is essentially that predicted by the amplitude (4.30), while the dotted curve is a polynomial $(A + B\cos\theta + C\cos^2\theta)$ fitted to the data at angles greater than 30°.

---

[41] Rough measurements at 265 Mev were made by Richter, Osborne, and Russell, *CERN Symposium on High Energy Accelerators and Pion Physics, Vol. 2*, p. 284 (1956). Results at 260 Mev were obtained by Knapp, Imhof, Kenney, and Perez-Mendez *Phys. Rev. 107*, 323 (1957).

[42] J. H. Malmberg and C. S. Robinson, *Phys. Rev. 109*, 158 (1958).

The turn up at small angles is a direct consequence of the pion current contribution. Agreement between theory and experiment is seen to be quite good. Similar results were obtained at 260 Mev by Knapp et al.[41]

The discussion of photoproduction given above is based on the static limit. There are relativistic modifications, such as a very small s-wave contribution to the neutral pion photoproduction amplitude, but the

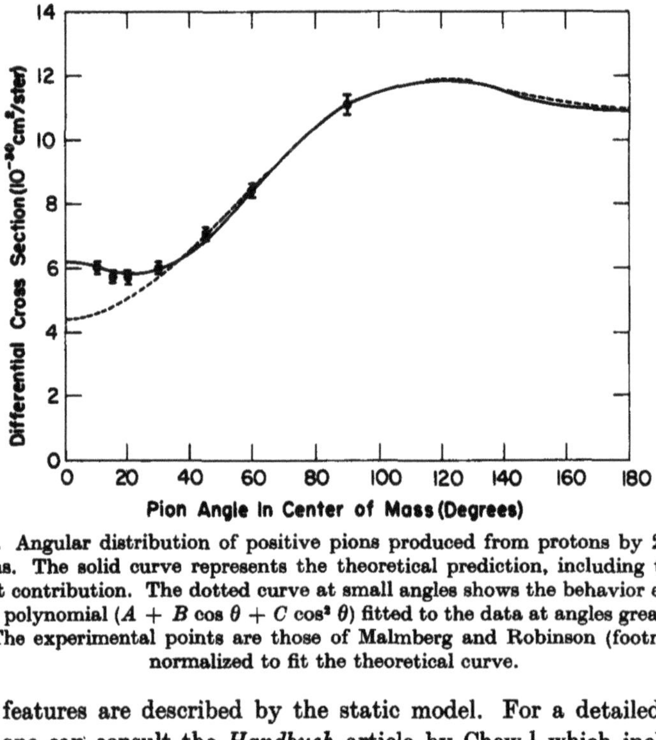

Fig. 7. Angular distribution of positive pions produced from protons by 225 Mev photons. The solid curve represents the theoretical prediction, including the pion current contribution. The dotted curve at small angles shows the behavior expected from a polynomial ($A + B \cos\theta + C \cos^2\theta$) fitted to the data at angles greater than 30°. The experimental points are those of Malmberg and Robinson (footnote 42), normalized to fit the theoretical curve.

gross features are described by the static model. For a detailed treatment one can consult the *Handbuch* article by Chew,[1] which includes a discussion of photoproduction by means of dispersion relations.[43] Koester and Goldwasser[44] have analyzed the data on angular distributions and total cross sections up to the resonance energy, and find that with a coupling constant $f^2 = 0.08$, the static theory plus $1/M$ corrections from the relativistic treatment is in reasonable agreement with all the data.

4.4. Anomalous Magnetic Moments of the Nucleons and Other Electromagnetic Phenomena

In the previous sections the problem of photoproduction of neutral and charged pions from nucleons has been discussed. The photoproduction

[43] Chew, Goldberger, Low, and Nambu, *Phys. Rev.* *106*, 1345 (1957).

[44] J. L. Koester, Jr. and E. L. Goldwasser, *Proceedings of the Seventh Annual Rochester Conference on High Energy Physics*, Interscience, New York, 1957, and private communication.

process involves high-frequency photons ($\omega > \mu$). But even at low frequencies the pion-nucleon interaction manifests itself in the electromagnetic properties of the nucleons themselves. For example, the proton and neutron are observed to have magnetic moments equal to $2.79e/2M$ and $-1.91e/2M$, respectively. Without the pion-nucleon interaction, we would expect the nucleons to act as "bare" Dirac particles, with the proton having a magnetic moment of $1.00e/2M$ and the neutron having none. The departure of the actual moment from the "bare" particle value is known as the anomalous magnetic moment $\Delta\mathscr{M}$:

$$\begin{aligned}\Delta\mathscr{M}_p &= +1.79\, e/2M \\ \Delta\mathscr{M}_n &= -1.91\, e/2M. \end{aligned} \tag{4.33}$$

The anomalous moments are attributed to the pion cloud which surrounds the nucleon. The fact that the anomalous moments are approximately equal and opposite can be understood in a simple way in terms of a charge-independent, pion-nucleon interaction.

If the spin and charge of the nucleon are considered fixed, the pion field satisfies (4.24), and for a point nucleon is given by (1.16). If the nucleon is represented by a source density $\rho(\vec{r})$, then the static field is:

$$\boldsymbol{\phi} = \frac{f}{\sqrt{4\pi}}\, \boldsymbol{\tau}\vec{\sigma} \cdot \vec{\nabla} Y(r), \tag{4.34}$$

where

$$Y(r) = \int \rho(\vec{r}')\, \frac{e^{-|\vec{r}-\vec{r}'|}}{|\vec{r} - \vec{r}'|}\, d\vec{r}'. \tag{4.35}$$

This pion field around the nucleon will contribute to the magnetic moment of the nucleon an amount $\Delta\mathscr{M}$:

$$\Delta\vec{\mathscr{M}} = \tfrac{1}{2} \int (\vec{r} \times \vec{j})\, d\vec{r}, \tag{4.36}$$

where $\vec{j}(\vec{r})$ is the current density due to the pion field (4.34). The current density is defined[45] by:

$$\vec{j} = ie(\phi_+ \vec{\nabla} \phi_- - \phi_- \vec{\nabla} \phi_+). \tag{4.37}$$

With the expression (4.34) for $\boldsymbol{\phi}$ it is easy to show that

$$\Delta\vec{\mathscr{M}} = e\tau_3 \frac{f^2}{4\pi} \int \vec{\sigma} \cdot \vec{\nabla} Y(r) \vec{L}\, \vec{\sigma} \cdot \vec{\nabla} Y(r)\, d\vec{r},$$

where $\vec{L} = \frac{1}{i}\vec{r} \times \vec{\nabla}$ is the orbital angular momentum operator. The dependence on τ_3 shows that the neutron and proton contributions are

[45] See G. Wentzel, footnote 6, p. 49.

equal and opposite. The "magnetic moment" is defined as the expectation value of $\Delta\vec{\mathscr{M}}$ in a state where the nucleon spin is up along the z-axis. This gives:

$$\Delta\mathscr{M} = e\tau_3 \frac{f^2}{4\pi}\frac{2}{3}\int \left[\frac{d}{dr}\,Y(r)\right]^2 d\vec{r}. \tag{4.38}$$

If we express $\Delta\mathscr{M}$ in units of $e/2M$ and introduce the Fourier transform $v(k)$ of the source density in (4.35), a little manipulation yields the result:

$$|\Delta\mathscr{M}| = \frac{8M}{3\pi} f^2 \int_0^\infty \frac{k^4|v(k)|^2}{(1+k^2)^2}\,dk. \tag{4.39}$$

With a point source the integral diverges. But we know from the discussion of the scattering that the source density, although rather compact, is still of finite extent. In fact, the scattering data of Section 3.1 can be used to specify the transform $v(k)$, once a shape is assumed, and so determine the anomalous magnetic moment (4.39). The relevant relations are (3.14) for the effective range r_α and the values of f^2 and ω_0^* from the effective range plot. With two shapes for $v(k)$, a square cut-off (1.7) and a Gaussian, fitted to the scattering data ($\omega_0^* = 2.1, f^2 = 0.08$ and 0.10), the anomalous magnetic moments resulting from (4.39) are tabulated in Table 3. The

Table 3. Anomalous Magnetic Moment in Nuclear Magnetons

Shape \ Coupling Constant	$f^2 = 0.08$	$f^2 = 0.10$
Square cut-off	1.76	1.55
Gaussian	1.67	1.52

values are seen to be very insensitive to the shape of the form factor $v(k)$ and only slightly more sensitive to changes in the coupling constant. The magnitude of the result is seen to be reasonably close to the experimental value (4.33). There are higher order corrections[46] which tend to increase the tabulated values by twenty-five or thirty percent. Consequently, the result given by (4.39) can be considered as satisfactory, especially since our treatment is oversimplified in that we have ignored nucleon recoil, heavy mesons, nucleon-antinucleon pairs, etc.

Other electromagnetic phenomena in which the pion-nucleon interaction enters importantly are the scattering of high-energy electrons by nucleons, and the low-energy interaction of neutrons with atomic electrons. These two problems are closely related and involve the electromagnetic structure of the nucleons. The structure can be described in terms of

[46] M. H. Friedman, *Phys. Rev.* 97, 1123 (1955).

spread out distributions of charge and magnetic moment with certain root mean square radii. It is customary to speak of a pion contribution and a "core" contribution due to the bare nucleon plus heavy mesons and nucleon pairs. Theoretical results based on the cut-off model give the correct qualitative behavior for the pion contribution to both the charge and magnetic moment densities, but the experimental data imply that the charge radius of the *core* is equal to the charge radius of the pion contribution. This seems difficult to understand in view of the fact that the dimension to be associated with the core is presumably the nucleon Compton wavelength (0.21×10^{-13} cm), whereas for the pion cloud it is $\frac{1}{\sqrt{2}}\mu^{-1} \simeq 1.0 \times 10^{-13}$ cm. Conversely, if the core is actually as large as implied by these results, one may wonder at the reason for the success of the simple cut-off model up to pion energies of several hundred Mev.[47]

[47] For a detailed discussion of the problems involved, see the review paper of Yennie, Levy, and Ravenhall, *Rev. Mod. Phys.*, *29*, 144 (1957).

PART II

K-MESONS AND HYPERONS

CHAPTER 5

Empirical Facts on K-Mesons and Hyperons

The discussion of K-mesons and hyperons must take rather different form from that given for the pion-nucleon problem because of the very different state of our knowledge on the subject. We are just beginning to amass sufficient data to warrant attempts at a detailed theory. What exists so far is an empirical description of the facts, with some rules and methods for codifying the facts and anticipating general features of as yet unobserved phenomena. For a discussion of the historical development of the subject, as well as a comprehensive review of experimental techniques and results, the reader is referred to review papers by Franzinetti and Morpurgo and by Dalitz.[48] The latest experimental information can be found in the Rochester Conference Proceedings.[49]

5.1. Hyperons

By now there have been observed seven particles of mass in the interval from the nucleonic mass up to (but not including) the deuteron mass. The word "particle" is used in the sense of a system having a definite mass, a lifetime long compared to strong interaction times ($\sim 10^{-23}$ sec), and well-defined electromagnetic properties—a system which is now most conveniently thought of as "elementary" (see, however, Section 7.1 below

Table 4. Basic Properties of Baryons

Particle	Mass (m_e)	Mass (Mev)	Mass (m_π)	Mean life (sec)
p	1836.1	938.2	6.72	– – – – –
n	1838.7	939.5	6.73	1050 ± 200
Λ^0	2181.7	1114.8 ± 0.2	7.99	2.9 ± 0.2 (−10)
Σ^0	2326.3	1189 ± 2.0	8.51	<1 (−11)
Σ^+	2328.3	1189.7 ± 0.3	8.51	0.7 ± 0.1 (−10)
Σ^-	2341.9	1196.7 ± 0.4	8.54	1.6 ± 0.2 (−10)
Ξ^-	2585.	1321 ± 4	9.46	$5 < T < 200$ (−10)
Ξ^0	?	?	?	?

[48] C. Franzinetti and G. Morpurgo, *Nuovo Cimento, VI,* Suppl. No. 2, 469 (1957); R. H. Dalitz, *Reports on Progress in Physics 20,* 163 (1957). For earlier reviews see H. S. Bridge and R. W. Thompson, *Progress in Cosmic Ray Physics,* Vol. III, ed. J. G. Wilson (North-Holland, Amsterdam, 1956).

[49] *Proceedings of Seventh Annual Rochester Conference on High Energy Physics* Interscience, N. Y., 1957.

for an alternative view). These seven particles are collectively known as "baryons", while the five which do not include the neutron and proton are called "hyperons". These particles are listed in Table 4, with their masses in various units, and their mean lifetimes in seconds.

An eighth particle, the neutral "cascade" particle, has been listed, although not observed. The reasons for its inclusion will be indicated below. The particles fall naturally into four mass groups or multiplets: (n,p), Λ^0, $(\Sigma^+, \Sigma^0, \Sigma^-)$, and (Ξ^-, Ξ^0)—and one can speak of two doublets, a singlet, and a triplet, as far as charges are concerned.

The decay processes of the various hyperons are:

$$\Lambda^0 \rightarrow \begin{cases} p + \pi^- + 37\text{ Mev} & (\sim 67\%) \\ n + \pi^0 + 40\text{ Mev} & (\sim 33\%) \end{cases} \tag{5.1}$$

$$\Sigma^0 \rightarrow \Lambda^0 + \gamma + 74\text{ Mev} \tag{5.2}$$

$$\Sigma^+ \rightarrow \begin{cases} p + \pi^0 + 116\text{ Mev} & (\sim 55\%) \\ n + \pi^+ + 110\text{ Mev} & (\sim 45\%) \end{cases} \tag{5.3}$$

$$\Sigma^- \rightarrow n + \pi^- + 117\text{ Mev} \tag{5.4}$$

$$\Xi^- \rightarrow \Lambda^0 + \pi^- + 67\text{ Mev} \tag{5.5}$$

The branching ratios for the Λ^0 and Σ^+ decay are only approximate, especially for the Λ^0 where the observation of the $(n + \pi^0)$ mode is difficult. We notice from Table 4 that the decays involving emission of pions all have comparable lifetimes, while the one known gamma mode is much shorter in lifetime. In fact, the longer lifetimes seem related in an elementary way to the volume in phase space; as more energy is available the lifetime is shortened. The factor of two between the Σ^+ and Σ^- lifetimes is plausible since there are two charge channels available to the Σ^+ compared to one for the Σ^-.

A spectroscopic term scheme for the baryons with the "energy levels" and decay modes indicated is shown in Fig. 8, solid lines being π emission modes, and dotted lines, gamma emission modes. The vertical scale is in units of the pion rest energy, but the splitting of the groups has been greatly exaggerated. This level scheme contains a number of puzzles concerning the properties of the hyperons. For example, why does the Σ^0 decay by gamma emission to the Λ^0, rather than to the neutron? On energetic grounds alone, dipole emission would be $(249/74)^3 = 38$ times more probable for the $\Sigma^0 \rightarrow n + \gamma$ transition than for $\Sigma^0 \rightarrow \Lambda^0 + \gamma$ transition. Similarly, why does the cascade particle not decay directly to the neutron, rather than "cascading" through the Λ^0? Clearly something other than energetics is involved. We can suspect that there is some kind of new selection rule operating here—a rule which must be added to the well known selection rules already assumed. Before discussing the selection rule, the K-mesons which are also involved in the ultimate scheme should be described.

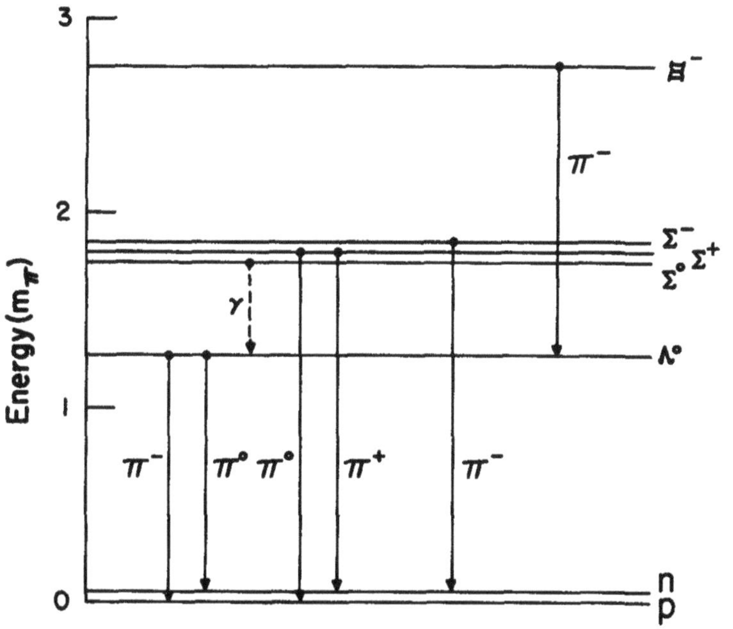

Fig. 8. Energy level diagram for baryons. The energies are expressed in units of the pion mass, but the splittings within the different multiplets has been exaggerated.

5.2. K-Mesons

At the same time as the hyperons were being discovered, indeed, often in the same experiment, particles were detected with masses lighter than a nucleon mass but much heavier than a pion mass. These were described as heavy mesons or K-mesons. When first observed, they represented a bewildering array of particles, charged and neutral, lying in the mass range 800–1100m_e. There are now reasons for believing (or at least hoping) that there is only one K-meson multiplet, consisting of a positive and neutral K-meson, together with their antiparticles. Empirically, the K-mesons appear as $K^{\pm}$, K^0. Their properties are summarized in Table 5.

Table 5. Properties of K-Mesons

Particle	Mass (m_e)	Mass (Mev)	Mass (m_π)	Mean Life (sec)
K^0	965.	493 ± 5	3.54	0.95 ± .08 (−10)
$K^{\pm}$	966.7	494.0 ± 0.1	3.54	1.24 ± .02 (−8)

The reason for the early bewilderment was the multiplicity of different decay modes. Each physically different mode of decay was phenomenologically designated as a different particle, and it was not until recently that it was established that all the "different" (charged) mesons had the same mass and lifetime. The various known decay processes are summarized below, with their phenomenological designations on the left.

$$\theta^0. \qquad K^0 \rightarrow \begin{cases} \pi^+ + \pi^- + 215 \text{ Mev} & (\sim 88\%) \\ \pi^0 + \pi^0 + 224 \text{ Mev} & (\sim 12\%) \end{cases} \tag{5.6}$$

$$\begin{matrix} \theta^+: \\ \tau: \\ \tau': \\ K_{\mu 2}: \\ K_{\mu 3}: \\ K_{e3}: \end{matrix} \qquad K^+ \rightarrow \begin{cases} \pi^+ + \pi^0 + 220 \text{ Mev} & (\sim 25\%) \\ \pi^+ + \pi^- + \pi^+ + 75 \text{ Mev} & (\sim 6\%) \\ \pi^+ + \pi^0 + \pi^0 + 84 \text{ Mev} & (\sim 2\%) \\ \mu^+ + \nu + 388 \text{ Mev} & (\sim 58\%) \\ \mu^+ + \pi^0 + \nu + 253 \text{ Mev} & (\sim 5\%) \\ e^+ + \pi^0 + \nu + 358 \text{ Mev} & (\sim 4\%) \end{cases} \tag{5.7}$$

There may well be other K^0 decay modes, but those in (5.6) are all that are known at present. The existence of $K^0 \rightarrow 2\pi^0$, observed by Eisler et al,[50] implies that the spin of the K^0 is even, since two identical bosons must be in a symmetrical state. The branching ratios in (5.6) and (5.7) are not very accurate, but show that for K^+ the dominant mode is $(\mu + \nu)$, with $(\pi + \pi)$ next in importance. This can be understood very qualitatively in terms of phase space arguments, but not in detail. The branching ratios for positive K-mesons seem to be independent of the mode of production, the same percentages being found from cosmic ray K-mesons as from machine produced ones. This strongly implies that the different decays are merely different channels for the decay of one particle. There are little data on K^- decays because K^- coming to rest in matter will be captured by a nucleon or nucleus before decaying, but it is assumed that K^- decays in the same way as K^+.

5.3. The Tau-Theta Puzzle

The idea that the various empirical particles (θ, τ, etc) are merely different manifestations of a single particle is very attractive and was seized upon early. It ran into difficulty very soon, a difficulty described as the "tau-theta puzzle". The puzzle is that, if we accept the standard conservation and symmetry laws, the tau meson ($\tau^\pm \rightarrow \pi^+ + \pi^- + \pi^\pm$) cannot have the same spin and parity as the theta meson ($\theta^\pm \rightarrow \pi^\pm + \pi^0$). As is now well known, this puzzle was the origin of the conjecture of Lee and Yang[51] that parity is not a valid concept in decay processes (see Part III). For the moment, however, let us apply the "old fashioned" ideas and assume that parity is conserved along with energy, linear and angular momentum.

Consider first the theta meson. Since it decays into two bosons, the

[50] Eisler, Plano, Samios, Schwartz, and Steinberger, *Nuovo Cimento*, *V*, 1700 (1957).
[51] T. D. Lee and C. N. Yang, *Phys. Rev.* *104*, 254 (1956).

parity of the final state is given by the relative orbital angular momentum of the two pions, which is the same as the total J of the system. If parity is conserved in the decay, the parity of the theta is $(-1)^J$, where J is its angular momentum.

For the tau meson we will summarize the spin-parity analysis of Dalitz.[52] In the decay into three pions, two are identical in charge, and one is different. Let the relative momentum of the identical pions be q, and the momentum of the odd pion relative to the center of gravity of the pair be p. For a positively charged tau meson the momenta are indicated in Fig. 9. The only relevant angle is the angle θ between the

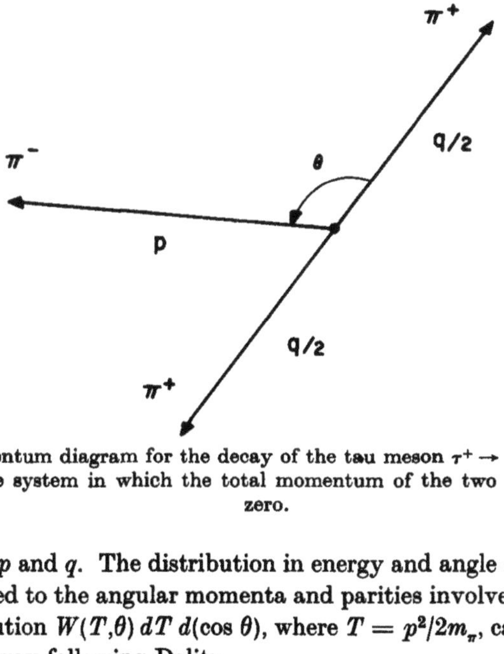

Fig. 9. Momentum diagram for the decay of the tau meson $\tau^+ \rightarrow \pi^+ + \pi^- + \pi^+$ in the coordinate system in which the total momentum of the two identical pions is zero.

direction of p and q. The distribution in energy and angle of the odd pion will be related to the angular momenta and parities involved in the decay. This distribution $W(T,\theta)\, dT\, d(\cos\theta)$, where $T = p^2/2m_\pi$, can be estimated in a simple way following Dalitz.

The pions can be treated nonrelativistically without too great error since the maximum energy of any pion is 50 Mev in the center of mass. Let L be the angular momentum quantum number of the relative motion of the identical pions, and ℓ the angular momentum quantum number of the odd pion relative to the center of mass of the pair. The total angular momentum of the tau meson is J. By the usual rules, J is limited to the range $|\ell - L| < J < \ell + L$. Since L is always even (2 identical bosons), the parity of the tau (actually of the final state) is $(-1)^{\ell+1}$ (remember the odd parity of the pion). Some possible spin-parity assignments are listed

[52] R. H. Dalitz, *Phil. Mag.* *44*, 1068 (1953), and *Phys. Rev.* *94*, 1046 (1954).

below, with the few lowest combinations (L,l) consistent with each given J and parity:

Spin-parity of τ	(L,l)	
0^+	--	
0^-	(0,0)	(2,2), . . .
1^+	(0,1)	(2,1), . . .
1^-	(2,2)	(4,4), . . .
2^+	(2,1)	(2,3), . . .
2^-	(0,2), (2,0)	(2,4), (4,2), . . .
3^+	(0,3), (2,1)	(2,3), (4,1), . . .
3^-	(2,2)	(2,4), (4,2), . . .

In each case the lowest values of (L,l) are given on the left of the dotted line, while possible higher combinations are indicated to the right.

The energy release is relatively small and the wavelengths of the pions are large compared to the decay interaction dimension which is presumably of the order of the tau meson Compton wavelength. Consequently the matrix elements for the decay will be determined solely by the centrifugal barrier penetration and will be of the form:

$$\langle f|H_{\text{int}}|i\rangle \sim \left(\frac{p}{Mc}\right)^l \left(\frac{q}{Mc}\right)^L, \tag{5.8}$$

where M is the K-meson mass. It is assumed that $p/Mc,\ q/Mc \ll 1$, so that only the lowest combinations of (l,L) values need be considered. From the list of possibilities it is clear that for 0^-, 1^+, 1^-, 2^+, 3^-, there is a unique prediction for $W(T,\theta)$:

Spin-parity	$W(T,\theta)$ $(T_0 = 50$ Mev$)$
0^-	$T^{1/2}(T_0 - T)^{1/2}$
1^+	$T^{3/2}(T_0 - T)^{1/2}$
1^-	$T^{5/2}(T_0 - T)^{5/2} \sin^2\theta \cos^2\theta$
2^+	$T^{3/2}(T_0 - T)^{5/2} \sin^2\theta$
3^-	$T^{5/2}(T_0 - T)^{5/2} \sin^2\theta\,(5 + 3\cos^2\theta)$

The experimental data[53] on τ^+ decays are consistent with an isotropic angular distribution and an energy dependence appropriate to 0^-, although 2^- and 3^+ (non-unique distributions) cannot be excluded. The important point is that spin-parity combinations with parity $= (-1)^J$ are definitely excluded. There are objections to the Dalitz analysis,[54] but general arguments show that if the parity is $(-1)^J$, there is a $(\sin^2\theta)$

[53] Baldo-Ceolin, Bonetti, Greening, Limentani, Merlin, and Vanderhaeghe, *Nuovo Cimento*, *VI*, 84 (1957).

[54] Eisenberg, Lomon, Rosendorff, *Nuovo Cimento*, *IV*, 610 (1956).

factor in the angular distribution. Such a factor seems inconsistent with the data.

We are thus faced with the tau-theta puzzle—the theta "must" have parity $(-1)^J$, and the tau "cannot" have parity $(-1)^J$. Various ingenious schemes[55] were invented to get around this difficulty, but they are of little interest now that violation of space inversion symmetry is known to occur in decay processes. This is not to say that the puzzle is completely solved, but rather that we must look in new directions and the old attempts are likely irrelevant. We will assume that the large variety of decay modes shown in (5.6) and (5.7) are in some unexplained way the result of the decay of a single (neutral or charged) K-meson, probably of spin 0 and unknown parity. This is consistent with the existence of the $2\pi^0$ mode for the neutral K-meson, as well as the Dalitz analysis of the tau decay.

5.4. Production versus Decay, Associated Production

In the discussion of the hyperons certain questions arose about the manner of their decay, implying that something curious was going on. The curious nature of the particles can be illustrated in another way by considering the problem of their relatively abundant production in high-energy collisions compared with their comparatively slow decay. In nucleon-nucleon collisions at a few Bev the hyperon production cross sections are about a tenth of a millbarn, and in pion-nucleon collisions are somewhat larger. These are appreciable cross sections. Now suppose that Λ^0's were produced in a reaction such as

$$\pi^- + p \rightarrow \Lambda^0 + \pi^0 \qquad \text{(strong)}. \tag{5.9}$$

Then one would expect a rapid decay via

$$\Lambda^0 \xrightarrow[\text{strong}]{} (\pi^0 + \pi^- + p) \xrightarrow[\text{strong}]{} p + \pi^-, \tag{5.10}$$

where the time scale would be of the order 10^{-20} sec or less. Instead, the decay time of Λ^0 is $\sim 10^{-10}$ sec. This is a fundamental problem in the understanding of these new particles. There have been a number of suggested explanations, such as very high spins for the hyperons,[56] but the most fruitful has been the idea of "associated production" due to Pais.[57] The hyperons and K-mesons are assumed to be produced always in pairs, a typical reaction being:

$$\pi^- + p \rightarrow (\pi^- + \Lambda^0 + K^0) \rightarrow \Lambda^0 + K^0 \quad \text{(strong)}, \tag{5.11}$$

---

[55] T. D. Lee and J. Orear, *Phys. Rev. 100*, 932 (1955); T. D. Lee and C. N. Yang, *Phys. Rev. 102*, 290 (1956).

[56] The high hyperon spin would hinder the decay process in which only a small amount of energy is released, but would not enter importantly in the production act if there is plenty of energy available. The spin values necessary, however, are of the order of 11/2 or greater, and seem unreasonable.

[57] A. Pais, *Phys. Rev. 86*, 663 (1952).

the strong interactions being such that the pion-nucleon and pion-hyperon interactions are of a similar nature, but that a nucleon is transformed into a hyperon and vice versa, by the emission or absorption of a K-meson. This assures that a reaction such as (5.9) cannot occur. On the other hand, in hyperon decay there is not enough energy to emit a K-meson. Consequently the decay is assumed to proceed via a weak interaction:

$$\Lambda^0 \rightarrow p + \pi^- \qquad \text{(weak)}. \tag{5.12}$$

The associated production idea of Pais has been strikingly confirmed by experiments on the Brookhaven Cosmotron and the Berkeley Bevatron. No clear cut example is known in which a hyperon is produced without an accompanying K-meson, whereas there are many examples of associated production. It should perhaps be noted, however, that the amount of evidence is not huge, and that the idea of associated production cannot be regarded as an absolute rule in the same sense as conservation of energy or angular momentum. It is, nevertheless, a very useful working rule for discussion of the production of hyperons and K-mesons.

CHAPTER 6

Classification Scheme and Details of the Interactions

Out of the peculiarities of the decay processes as summarized in Fig. 8 and the idea of associated production came the concept of a new quantum number, generally called "strangeness", and a classification scheme for the baryons and mesons due to Gell-Mann[58] and Nishijima.[59] This classification scheme is not a theory, but rather a set of rules to allow the correlation of the empirical facts on K-mesons and hyperons. We will first discuss the classification of the particles and then examine some of the details of the strong and weak interactions in the light of this classification.

6.1. Displaced Charge Multiplets

The emergence of some sort of order out of the seven or eight baryons and the K and π mesons starts with the recognition of *four* different categories of particles:

(1) Baryons—heavy fermions, nucleons and hyperons

(2) Mesons—pions and K-mesons

(3) Photons

(4) Leptons—light fermions, muons, electrons, neutrinos,

and *three* basic types of interactions:

(a) Strong—baryon-meson interactions

(b) Electromagnetic interactions

(c) Weak—all decay interactions.

Then one notes that the baryons fall naturally into mass multiplets, as can be seen from Table 4 or Fig. 8. Since it is the strong interactions which presumably determine the gross structure of the mass spectrum it is natural to assume that *all strong interactions are charge-independent* (as is the pion-nucleon interaction), and that the mass multiplets can be thought

---

[58] M. Gell-Mann, *Phys. Rev.* *92*, 833 (1953), *Nuovo Cimento*, *IV*, Series X, Suppl. No. 2, 848 (1956).

[59] T. Nakano and K. Nishijima, *Prog. Theo. Phys.* *10*, 581 (1953); K. Nishijima, *Fort. der Physik*, *IV*, 519 (1956).

of as *isotopic spin multiplets*. The assignments of isotopic spin are as follows:

$$\begin{array}{ll} n,\, p & T = \tfrac{1}{2} \\ \Lambda^0 & T = 0 \\ \Sigma^-,\, \Sigma^0,\, \Sigma^+ & T = 1 \\ \Xi^-,\, (\Xi^0) & T = \tfrac{1}{2} \end{array} \tag{6.1}$$

The last assignment is, for the moment, hypothetical. Within the Gell-Mann-Nishijima scheme it is necessary for a natural explanation of the decay of Ξ^-, as will be described below.

As soon as we include electromagnetic interactions the connection between the z-component of isotopic spin and the charge arises. For nucleons the relation between charge and isotopic spin is:

$$T_z = \begin{Bmatrix} +\tfrac{1}{2} & p \\ -\tfrac{1}{2} & n \end{Bmatrix} \quad Q = T_z + \tfrac{1}{2}, \tag{6.2}$$

where Q is the charge of the nucleon. For antinucleons, the connection is

$$T_z = \begin{Bmatrix} -\tfrac{1}{2} & \bar{p} \\ +\tfrac{1}{2} & \bar{n} \end{Bmatrix} \quad Q = T_z - \tfrac{1}{2}. \tag{6.3}$$

For the pions

$$T_z = \begin{Bmatrix} +1 & \pi^+ \\ 0 & \pi^0 \\ -1 & \pi^- \end{Bmatrix} \quad Q = T_z\,. \tag{6.4}$$

We see that the charge (an absolutely conserved quantity) can be related to the z-component of isotopic spin for nucleons, antinucleons and pions by means of

$$Q = T_z + \frac{n}{2}, \quad \text{where } n = \begin{cases} 1 \text{ for nucleons} \\ -1 \text{ for antinucleons} \\ 0 \text{ for pions.} \end{cases} \tag{6.5}$$

The n is recognized as the "baryon number", another absolutely conserved quantity.[60] For an assembly of pions and nucleons (6.5) has an obvious generalization.

For the hyperons and K-mesons the relation between charge and z-component of isotopic spin is not given by (6.5), but is given by the generalization:

$$Q = T_z + \frac{n}{2} + \frac{S}{2}, \tag{6.6}$$

[60] The baryon or heavy particle number is the number of particles minus the number of antiparticles. There is no known process in which the baryon number changes, even though the number of particles or the number of antiparticles may change individually.

where n is the baryon number, and S is a new quantum number, the strangeness. The shifting of Q relative to T_z by means of S gives rise to the name "displaced charge multiplets".

Q and n are conserved in all processes, and T_z is conserved in the strong and electromagnetic interactions. Consequently, S is conserved in strong and electromagnetic interactions. Ordinary nucleons and pions have $S = 0$, but the hyperons and K-mesons have $S \neq 0$. The stability of these particles against decay into ordinary ones now appears as a "consequence" of $\Delta S \neq 0$ which can occur only for the very weak decay interactions.

The assignment of the quantum number S is straightforward:
(i) $\boldsymbol{\Lambda^0}$. This particle forms an isotopic spin singlet ($T = 0$, $T_z = 0$). Since it is produced in a reaction involving nucleons and pions or nucleons, it must be a "particle" with $n = +1$, not an "antiparticle". From (6.6) one finds $S = -1$.
(ii) $\boldsymbol{\Sigma^-}$, $\boldsymbol{\Sigma^0}$, $\boldsymbol{\Sigma^+}$. These particles are an isotopic spin triplet ($T = 1$, $T_z = -1, 0, 1$). As for the lambda, $n = +1$, and consequently $S = -1$. We note that the observed neutral sigma decay mode $\Sigma^0 \rightarrow \Lambda^0 + \gamma$ has $\Delta S = 0$, and so will be very rapid in agreement with observation.
(iii) $\boldsymbol{\Xi^-}$. The decay process $\Xi^- \rightarrow \Lambda^0 + \pi^-$ is a typically slow decay, while the decay $\Xi^- \rightarrow n + \pi^-$ is not observed. These facts exclude $S = 0, -1$ as possible assignments of strangeness. If the Ξ^- is an isotopic spin singlet, (6.6) demands $S = -3$. Gell-Mann favors the assignment $S = -2$, implying an isotopic spin doublet (Ξ^-, Ξ^0), although the neutral member is as yet unobserved. The assignment of $S = -2$ is equivalent to the postulate that $\Delta S = \pm 1$ for decay processes. This is consistent with all known hyperon and K-meson decay modes.
(iv) **K-mesons.** According to (5.6) and (5.7), both neutral and charged K-mesons decay into ordinary particles. Assuming that $\Delta S = \pm 1$ in the decay, it is necessary to assign $S = +1$ or $S = -1$ to the K-mesons. For $S = +1$, (6.6) implies that $T = \frac{1}{2}$, $T_z = \pm\frac{1}{2}$, corresponding to K^+ and K^0 mesons. It is assumed that for antiparticles the strangeness has the opposite sign, just as do T_z and n. Consequently there should be antihyperons with strangeness $S = +1, +2$. For the K-mesons there are known K^-, which fall in the category (K^-, $\bar{K}^0$) with $S = -1$.

The displaced charge multiplets can be shown graphically in an energy level diagram such as Fig. 10, where the masses of the strongly interacting particles are plotted as a function of strangeness. Only the antiparticles of the K-mesons are shown in Fig. 10, since they are the only ones as yet observed, aside from the antinucleons. The baryons within a given vertical column (same strangeness) interact strongly only through the pion field (of zero strangeness). On the other hand, baryons in one column can interact with baryons in the column on either side ($\Delta S = \pm 1$) only through the appropriate K-meson field to conserve the total strangeness of the system.

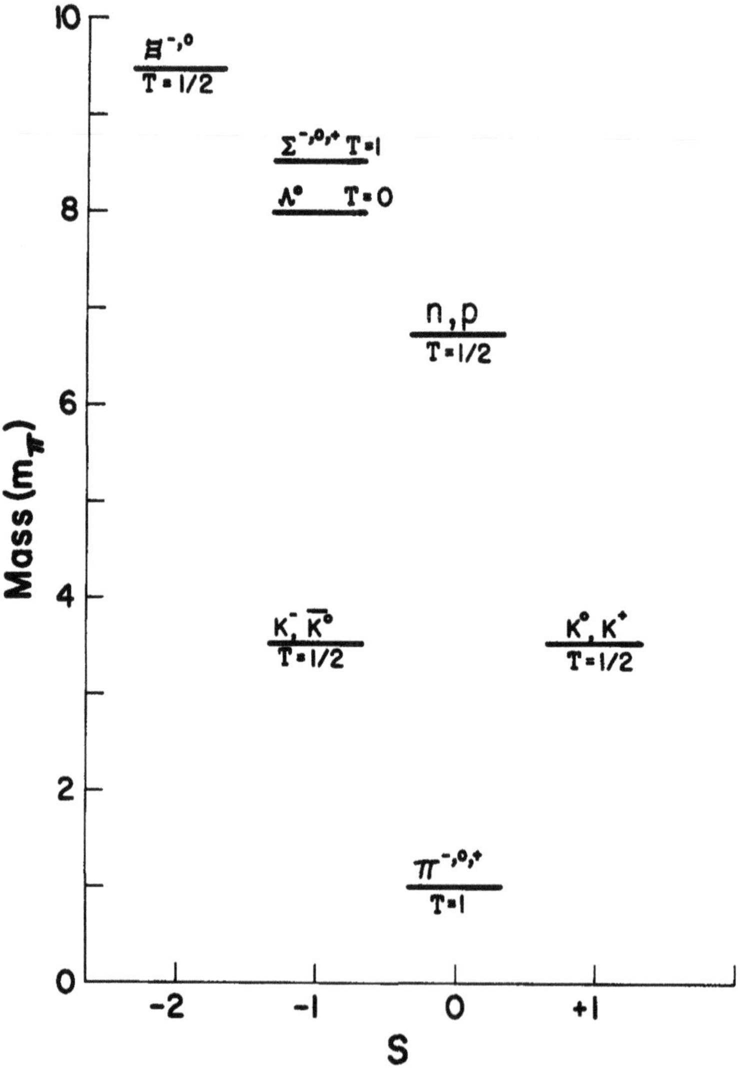

Fig. 10. Energy level diagram of the displaced charge multiplets as a function of the strangeness quantum number S. The particle masses are in units of the pion mass. Only the antiparticles of the (K^+, K^0) doublet are specifically shown, but it is assumed that there are antiparticles for the hyperons as well, with opposite strangeness.

There are a number of different ways of phrasing the "strangeness" concept,[61] but they are all essentially equivalent to the Gell-Mann-Nishijima scheme.

6.2. Production and Capture Processes, Hyperon Spins

Pais' idea of associated production is, of course, implicit in the classification scheme of Section 6.1, since S is assumed to be conserved in the strong interactions. If, for example, a Λ^0 (with $S = -1$) is produced in a collision involving nucleons with pions or nucleons, it must be accompanied by either K^+ or K^0 ($S = +1$) in order to conserve strangeness in the production act. The following list gives some examples of production reactions which are allowed, and a few examples which are forbidden by the assignments of strangeness made above. The letter N stands for a nucleon; bars above letters denote antiparticles; approximate thresholds for stationary nucleon targets are indicated; charges are not explicitly shown:

Some Allowed Reactions		Threshold (Bev)
$\pi + N \rightarrow$	$\Lambda + K$	0.76
	$\Sigma + K$	0.90
	$N + K + \bar{K}$	1.36
	$\Xi + K + K$	2.22
$N + N \rightarrow$	$N + \Lambda + K$	1.59
	$N + \Sigma + K$	1.79
	$N + N + K + \bar{K}$	2.49
	$N + \Xi + K + K$	3.75

(6.7)

Some Forbidden Reactions

$$\pi + N \rightarrow \begin{cases} \Lambda + \bar{K} \\ \Sigma + \bar{K} \\ \Lambda + K + \bar{K} \\ \Xi + K \end{cases} \qquad N + N \rightarrow \begin{cases} \Lambda + \Lambda \\ \Lambda + \Sigma \\ \Lambda + \Xi \end{cases} \tag{6.8}$$

It is easy to see that the production reactions with the lowest thresholds are those involving the production of K-mesons with $S = 1$, denoted by the symbol K. In order to produce K-mesons with $S = -1(\bar{K})$, it is necessary to make a pair of K-mesons $(K + \bar{K})$. This has a higher threshold than the single production and can be done in fewer ways. Consequently it is expected that the ratio of K^+/K^- from machines will be

[61] R. G. Sachs, *Phys. Rev. 99*, 1573 (1955); A. Salam and J. C. Polkinghorne, *Nuovo Cimento, II*, 685 (1955); J. Schwinger, *Phys. Rev. 104*, 1164 (1956).

large, at least at energies not too far above the thresholds. This is, in fact, the observed result, the ratio K^+/K^- being of the order of one hundred for protons of a few Bev energy on complex nuclei, and of the order of five to ten for pions of similar energies on nuclei.

As was mentioned in Section 5.4, no evidence exists for production reactions such as (6.8) which do not conserve strangeness, while there is ample evidence for many of the reactions in (6.7).

There are very few quantitative data on the production reactions as yet, but some work has been done with the first two processes in (6.7). The total cross sections observed are of the order of a few tenths of a millibarn for incident pions with energies from threshold to 1.3 Bev. The weak energy dependence of these cross sections, as well as their magnitude, can be understood qualitatively in terms of Fermi's statistical model for high energy collisions.[62] The angular distributions for the four reactions

$$\pi^- + p \rightarrow \begin{cases} \Lambda^0 + K^0 \\ \Sigma^0 + K^0 \\ \Sigma^- + K^+ \end{cases} \qquad \pi^+ + p \rightarrow \Sigma^+ + K^+$$

have been studied for pions with kinetic energies of 1.1 to 1.3 Bev. A marked dependence on angle is observed, with the charged hyperons peaked forwards in the center of mass system and the neutral hyperons peaked backwards. If the reactions are interpreted as proceeding through states of definite isotopic spin, the angular distributions imply that a strong mixture of $T = \frac{1}{2}$ and $T = \frac{3}{2}$ states is involved in the Σ production. In Λ production only the $T = \frac{1}{2}$ state can appear.

For nucleon-nucleon collisions the meager data indicate a production cross section of a few tenths of a millibarn at 3 Bev.

The capture of negative K-mesons in hydrogen gives a further illustration of the concept of strangeness; it also yields information on the K^--meson-nucleon interaction and the spins of the hyperons. The capture reactions observed with low-energy K^- stopping in a hydrogen bubble chamber[63] are:

$$K^- + p \rightarrow \Lambda^0 + \pi^0 + 183 \text{ Mev} \tag{6.9}$$

$$K^- + p \rightarrow \begin{cases} \Sigma^- + \pi^+ + 96 \text{ Mev} & (\sim 50\%) \\ \Sigma^0 + \pi^0 + 109 \text{ Mev} & (\sim 25\%) \\ \Sigma^+ + \pi^- + 103 \text{ Mev} & (\sim 25\%) \end{cases} \tag{6.10}$$

Reaction (6.9) occurs about $\frac{1}{3}$ as frequently as the neutral sigma production in (6.10). If isotopic spin is conserved in the capture process, the

[62] R. Serber, footnote 49.

[63] Alvarez, Bradner, Falk-Vairant, Gow, Rosenfeld, Solmitz, and Tripp, *Nuovo Cimento*, *V*, 1026 (1957).

branching ratios in (6.10) imply that a coherent mixture of $T = 0$ and $T = 1$ states occurs.[64]

Other capture reactions are possible: for example, the deuterium capture processes:

$$K^- + d \rightarrow \begin{cases} \Sigma^- + n + \pi^+ + 94 \text{ Mev} \\ \Lambda^0 + p + \pi^- + 176 \text{ Mev.} \end{cases} \tag{6.11}$$

These are of particular interest because of the possibility of appreciable bound state formation between hyperon and nucleon[65] (if the forces are attractive). Such systems would provide fundamental information on hyperon-nucleon forces (see Section 6.4).

The angular distributions of the hyperon decay products in both the production process ($\pi^- + p \rightarrow H + K$) and the capture at rest ($K^- + p \rightarrow H + \pi$) can be used to give information on the hyperon spins.[66] In the production act the argument depends on conservation of angular momentum along the direction defined by the incident pion for those hyperons which are produced in the forward or backward direction. Since the incident pion and outgoing K-meson, both assumed to have spin zero, have no component of orbital angular momentum along the incident direction, only the magnetic substates $m = \pm\frac{1}{2}$ of those hyperons' spins can be populated. In the subsequent decay of the hyperon there will be an anisotropic angular distribution of decay products if the spin is greater than $\frac{1}{2}$. For example, for spin $\frac{3}{2}$ the distribution is $(1 + 3\cos^2\theta)$. Fortunately the hyperons are produced predominantly in the forward and backward directions, and a large fraction of the total are useful for such an analysis. Experimental data* for Λ^0 and Σ^- hyperons produced forwards or backwards at pion energies from 0.9 to 1.3 Bev indicate an isotropic angular distribution for the decay products, implying spin $\frac{1}{2}$ for the hyperons assuming the K-meson spin is zero.

In the capture reactions (6.9) and (6.10) the same arguments about magnetic substates and decay product angular distributions relative to the hyperon line of flight apply, provided the capture is mainly from s-states around the proton. Data from Berkeley[63] on the decay of Σ's in the reaction (6.10) are consistent with spin $\frac{1}{2}$ for the Σ. Possible contributions from capture in p-states, however, make this evidence somewhat less convincing than the production results.

---

* Eisler et al, *Nuovo Cimento, VII*, 222 (1958).

[64] For a pure $T = 0$ state the ratios are 1:1:1, while for a pure $T = 1$ state they are 1:0:1. If there is an *incoherent mixture* of $T = 0$ and $T = 1$ states, the Σ^+ and Σ^- branches should have equal intensities given by an appropriate weighting of the two extremes, provided small effects such as differences in energy release can be ignored.

[65] A. Pais and S. B. Treiman, *Phys. Rev. 107*, 1396 (1957).

[66] The possibility of a spin determination in the production process was pointed out by R. K. Adair, *Phys. Rev. 100*, 1540 (1955); and for the K^--capture from bound states by S. B. Treiman, *Phys. Rev. 101*, 1216 (1956).

6.3. K-Meson Scattering and Absorption by Nucleons and Nuclei

The scattering of K-mesons by nuclei in photographic emulsions and by protons in hydrogen bubble chambers provides information on the K-nucleon interaction. For K^+ scattering, there are just the ordinary and charge exchange scatterings.

$$\begin{aligned} K^+ + N &\rightarrow K^+ + N \\ K^+ + n &\rightarrow K^0 + p \end{aligned} \tag{6.12}$$

But for K^- scattering there is the absorption in flight given by (6.9) and (6.10), as well as the ordinary and charge exchange scattering.

For K^+-mesons on hydrogen the total cross section is about 15 mb, independent of energy from 10 to 200 Mev, with a roughly isotropic angular distribution.[67] For K^+ scattering in emulsions the average cross section is ~10 mb/nucleon, independent of energy. The scattering from the neutrons in nuclei can give rise to exchange scattering. Experimentally the ratio of exchange to ordinary scattering in emulsions is of the order of 0.2. With the assumptions that isotopic spin is conserved and that the neutrons and protons scatter incoherently, one can show that this ratio is consistent with a K^+-nucleon interaction in a pure $T = 1$ state. We note, however, that the branching ratios in the K^- capture process (6.10) seem to exclude a pure $T = 1$ interaction between K^--mesons and nucleons. Furthermore, the K^+ exchange to ordinary scattering ratio seems to increase markedly at higher energies (~400 Mev).

The small angle scattering of K^+ in emulsions shows interference with the Coulomb scattering, the sign being such that *the K^+-nucleus interaction is repulsive*. A fit by means of the optical model can be made[68] with a complex potential having the same radius and diffuseness as found in electron scattering by nuclei,[69] a repulsive real part $V = 27 \pm 5$ Mev, and an imaginary part $W = 5.7 \pm 1$ Mev for the energy range 40–100 Mev and $W = 10 \pm 2$ Mev at 150 Mev. The energy variation of W, after allowance for the Pauli principle in collisions inside the nucleus, corresponds to a K^+-single nucleon cross section of about 15 millibarns independent of energy, in accord with the free particle cross section.

For negative K-mesons the absorption processes play an important role, especially at low energies. For K^--mesons in hydrogen a partial summary of the available data* is given in Table 6.

The ordinary scattering cross section has a roughly isotropic angular

---

* Ascoli, Hill, and Yoon, *Nuovo Cimento* (to be published).

[67] Lannutti, S. Goldhaber, G. Goldhaber, Chupp, Giambuzzi, Marchi, Quareni, and Wataghin, *Phys. Rev. 109*, 2121 (1958).

[68] Igo, Ravenhall, Tiemann, Chupp, G. Goldhaber, S. Goldhaber, Lannutti, and Thaler, *Phys. Rev. 109*, 2133 (1958).

[69] Hahn, Ravenhall, and Hofstadter, *Phys. Rev. 101*, 1131 (1956).

Table 6. Interactions of Negative *K*-Mesons in Hydrogen

Energy interval (Mev)	Mean energy (Mev)	Total cross sections (millibarns) Ordinary scattering	Reaction ($\Sigma^\pm$ only)
5–45	25	56 ± 13	36 ± 10
45–85	65	41 ± 8	12 ± 4
85–125	105	$23 \pm {}^{18}_{7}$	$12 \pm {}^{16}_{4}$

distribution and falls smoothly with increasing energy from a zero energy value of about 70 mb, while the reaction cross section behaves as a $\frac{1}{v}$ law or even steeper. The optical theorem (see (3.23)) can be used to show that both $T = 0$ and $T = 1$ interactions seem to be present, as was indicated for the reactions at rest (6.10). The behavior of the cross sections strongly imply that *s*-state interactions are dominant at these low energies.

In emulsions the absorption greatly dominates the ordinary scattering, being $\sim$20 times as large, at least for heavy nuclei. An optical model approach using Monte Carlo methods shows that the *K^--nucleus interaction is attractive*[70] ($V \simeq -25 \pm 10$ Mev). Similar results are obtained from small angle scattering through Coulomb interference.

We note that the zero energy K^--proton elastic scattering cross section is of the order of four times that of K^+-proton scattering. This is qualitatively what we should expect if the K^+-nucleon interaction is repulsive, while the K^--nucleon interaction is attractive. A simple model which describes the *K*-meson-nucleon interaction by a potential well can be used to correlate these data (see Section 7.1).

6.4. Hypernuclei

The replacement of an ordinary nucleon in a nucleus by a hyperon can lead to the formation of a metastable system with a lifetime of the order of the decay time of the hyperon. These are called "hyperfragments" or "hypernuclei." Not all hyperons will form metastable systems—for example, $\Sigma^- + p \rightarrow \Lambda^0 + n$ is a fast reaction. Only the Λ^0 can be added more or less indiscriminately, but under exceptional conditions others can be used ($\Sigma^+ + p$, $\Sigma^- + n$). We shall restrict our considerations to hypernuclei with one Λ^0 hyperon present.

Hypernuclei yield information on hyperon-nucleon forces. Although unstable, they exist for a very long time on a nuclear time scale, and their dynamics can be discussed in terms analagous to those employed in ordinary nuclear physics.[71]

Hypernuclei with Λ^0 replacing a nucleon have been observed, as

[70] Alles, Biswas, Ceccarelli, and Crussard, *Nuovo Cimento, VI*, 571 (1957).

[71] R. H. Dalitz, *Phys. Rev. 99*, 1475 (1955); *Proceedings of Sixth Annual Rochester Conference on High Energy Physics*, Interscience, New York, 1956.

fragments from a high-energy collision, chiefly in emulsions, with masses from $A = 3$ to 16. From measurements on the decay products (nucleons, with or without a pion) it is possible to find the binding energy of the Λ^0 in the nucleus. Typical binding energy values in Mev for light hypernuclei[72] are H^{2*} : ~0, H^{3*} : 0.25 ± 0.2, H^{4*} : 1.44 ± 0.25, He^{4*} : 1.7 ± 0.2, He^{5*} : 2.6 ± 0.2. We recall that for nucleons, aside from "shell structure" fluctuations, there is a tendency for saturation of the binding energy per nucleon because of the Pauli principle and the nature of the forces. For the hyperons the circumstances are quite different because the Pauli principle does not operate. Consequently, there should be a general increase in binding energy, roughly as the mass number of the nucleus, without saturation, at least for light nuclei. Fig. 11 illustrates

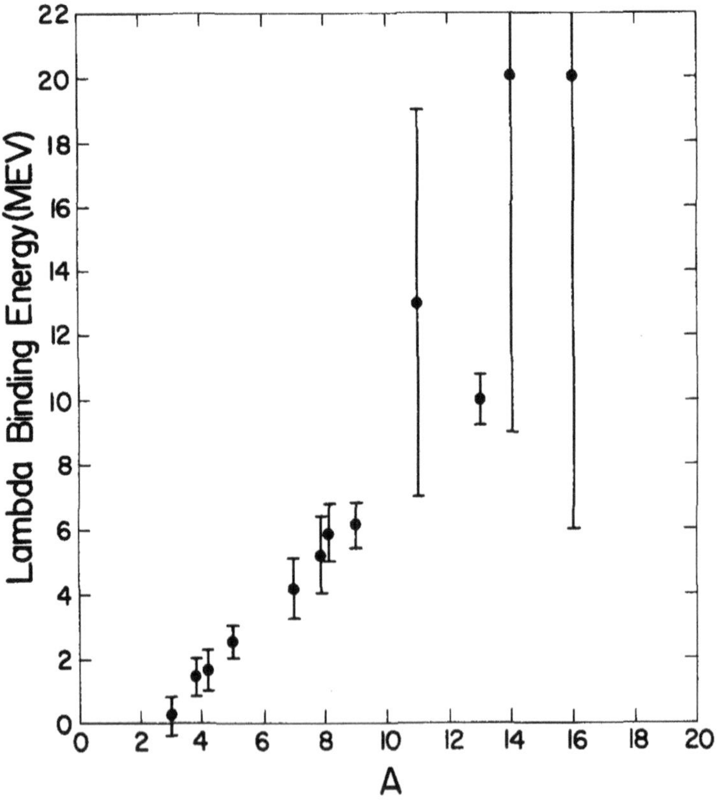

Fig. 11. Binding energy of Λ^0 hyperons in hypernuclei as a function of mass number. Two points together at a given A value correspond to hypernuclei of different charge, e.g. for $A = 4$, H^{4*} and He^{4*}. The binding energy for a given A represents an average over a number of observations. For $A > 10$ the identification of the hypernuclei and the binding energy is uncertain.

[72] The notation employed is Z^{A*} to describe a nucleus of mass number A with Z protons and a Λ^0 replacing a neutron.

just this feature for the Λ^0 binding energy as a function of A. The data are taken from the proceedings of the Seventh Annual Rochester Conference on High Energy Physics.[49]

In accord with our assumptions of charge independence in the baryon-meson interactions, we can assume that the $\Lambda^0 - N$ forces will be charge independent. Then the light hypernuclei can be interpreted in terms of isotopic spin multiplets, in analogy with ordinary light nuclei. A characteristic difference occurs, however, since the Λ^0 hyperon has $T = 0$, whereas all nucleons have $T = \frac{1}{2}$. This leads to a set of lowest multiplets illustrated in Fig. 12 where a schematic comparison is made between the

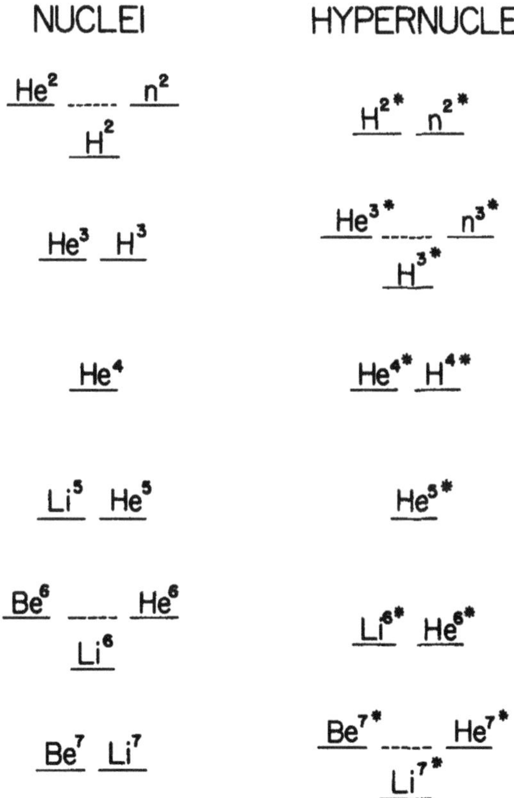

Fig. 12. Schematic diagram of low-lying isotopic spin multiplets for ordinary nuclei and their corresponding isobaric hypernuclei for $A = 2, \ldots, 8$. The differences between multiplets for a given mass number reflect the $T = 0$ isotopic spin of the Λ^0 hyperon.

familiar multiplets of ordinary nuclei and the corresponding isobaric hypernuclei. In general only dynamically stable ordinary nuclei are shown, although exceptions appear at $A = 2, 5$, etc. It will be noted that the multiplet structure in ordinary nuclei of mass number A appears for hypernuclei at mass number $(A + 1)$, as expected for a $T = 0$ hyperon.

The observation of both He^{4*} and H^{4*} nuclei with comparable binding energies, as well as Be^{8*} and Li^{8*} (see Fig. 11), confirms the assignment of $T = 0$ for the Λ^0 hyperon and the concept of charge independent $\Lambda^0 - N$ forces. The detection of a number of H^{3*} nuclei, but no He^{3*} nuclei, indicates that the ordering of levels in the level scheme shown in Fig. 12 for $A = 3$ is correct, and that the $T = 0$ state lies lowest.

Just as for the $A = 2$ multiplets of ordinary nuclei, this conclusion about the $A = 3$ hypernuclei shows us that the $\Lambda^0 - N$ forces are spin dependent. For He^{3*} or n^{3*} the nucleon spins must be opposed, and the Λ^0 will see no net nucleon spin. In H^{3*}, however, the nucleon spins can be either parallel or antiparallel. The fact that H^{3*} lies lowest implies that the $\Lambda^0 - N$ forces are spin dependent, and that the two nucleons' spins are aligned in the favored orientation.

Dalitz[71] has made a quantitative analysis of the binding energies of light hypernuclei and obtained information on the spin dependence of the forces. Since the $\Lambda^0 - N$ force arises from the exchange of virtual K-mesons or two or more pions (see Section 7.3), it will have a short range compared to ordinary nuclear forces. Consequently, the spatial dependence of the interaction can be described to a good approximation by a spin dependent potential:

$$V(\vec{r}) = \Omega \rho(\vec{r}), \tag{6.13}$$

where Ω is the volume integral of the $\Lambda^0 - N$ potential (in Mev-cm^3), and $\rho(\vec{r})$ is the effective nucleon density in the nucleus.[73] Dalitz takes the nucleon density from electron scattering data,[69] convoluted with a short range potential, and determines Ω for various hypernuclei in terms of their observed binding energies. The hypernucleus He^{5*} is the corner stone of the analysis, the assumption being that the very stable alpha particle provides a particularly suitable core which will be distorted little by the presence of the hyperon. The spin-averaged volume integral per nucleon found in this way is $\Omega_{av} \simeq 220.$ Mev $\times 10^{-39}$ cm^3, showing that the strength of the $\Lambda^0 - N$ forces is comparable with the strength of ordinary nuclear forces. Study of other light hypernuclei with an odd nucleon present shows that in the favored spin state of the $\Lambda^0 - N$ system the volume integral is $\Omega_{fav} \simeq 350.$ Mev $\times 10^{-39}$ cm^3. This indicates a

[73] This approximation assumes that the Λ^0 does not influence the distribution of nucleons appreciably. It therefore works best when the Λ^0 is weakly bound and spends most of its time outside the nucleus.

very strong spin dependence for the force. There is some evidence[74] that the force is strongest in the singlet spin state, but reliability of the arguments is as yet not clear. More refined calculations, as well as other information, must be provided in order to elucidate such details of the $\Lambda^0 - N$ interaction.

6.5. Spin of the Lambda from Hypernuclear Decay

The manner of disintegration of hypernuclei yields important evidence about the spin of the Λ^0 hyperon. The free Λ^0 always decays into a pion and a nucleon. In contrast, hypernuclei disintegrate most often into nucleons alone. Only the lightest ones sometimes decay with the emission of a pion as well as nucleons.

The mesonic and non-mesonic modes of decay can be described in terms very like those used to discuss the de-excitation of a nucleus by either gamma emission or internal conversion.[75] Just as the orbital electrons around an excited nucleus provide an additional mechanism for decay by direct Coulomb interaction, the nucleons in the neighborhood of a hyperon provide the mechanism for non-mesonic decay via their static pion fields. We know that the ordinary electromagnetic internal conversion coefficient is a very sensitive function of the spin change occurring in the nuclear de-excitation. Accordingly we would expect the relative abundances of the two hypernuclear decay modes to yield information about the hyperon spin.

In all but the lightest hypernuclei the simple analogy with internal conversion is complicated by two effects: (a) The mesonic decay of a bound hyperon is hindered by the effect of the Pauli exclusion principle on the nucleon of about 5 Mev emitted in the decay; (b) the real pion emitted in the mesonic decay of a bound Λ^0 may be absorbed before escaping from the nucleus. Both of these effects tend to make the non-mesonic decay mode dominate for heavier hypernuclei, in accord with observation. To gain relatively unambiguous information from the internal conversion process we need to consider the lightest hypernuclei ($Z = 1$, 2, and perhaps 3).

The free Λ^0 decay mode $\Lambda^0 \rightarrow p + \pi^-$ can be described phenomenologically by a decay interaction coupling the baryons to the pion field:

$$H_{\text{decay}} = \int \rho_{\text{eff}}\, \phi_-, \tag{6.14}$$

where $\rho_{\text{eff}} = \text{const.}\ (\psi_p^* \mathcal{O} \psi_\Lambda)$ is an effective transitional "charge" density, $\mathcal{O}$ being some unknown, perhaps non-local interaction operator responsible for the decay process. If the spin of the Λ^0 is such that the pion is emitted

[74] R. H. Dalitz, unpublished lectures at Brookhaven National Laboratory, July–August, 1957. See also R. H. Dalitz and B. W. Downs, *Phys. Rev.* (to be published.)

[75] W. Cheston and H. Primakoff, *Phys. Rev.* *92*, 1537 (1953).

with orbital angular momentum ℓ, the operative part of the pion field ϕ_- is (see (1.17)):

$$\phi_- = 4\pi(-i)^\ell \frac{q^\ell Y_\ell^m(\Omega_q)}{\sqrt{2\omega_q}} \times \frac{r^\ell Y_\ell^{*m}(\Omega_r)}{(2\ell+1)!!}, \tag{6.15}$$

where it has been assumed that the interaction is confined to distances small compared to the pion wavelength. A simple perturbation calculation based on (6.14) and (6.15) leads to the decay rate:

$$\Gamma_\pi = \lambda^2 q^{2\ell+1}, \tag{6.16}$$

where

$$\lambda = \frac{\sqrt{2}}{(2\ell+1)!!} \langle r^\ell Y_\ell^m \rho_{\text{eff}} \rangle \tag{6.17}$$

is the effective multipole moment for the transition. With 37 Mev energy release the pion momentum is $q = 0.72$ in units of μc.

For a hyperon bound in an environment of nucleons an additional decay mechanism exists, since the hyperon's effective "charge" density can interact with the *static* pion fields of the nucleons, rather than the virtual "radiation" field of (6.14). This will give rise to an effective interaction between the decaying hyperon and the nearby nucleons. If we consider one proton at coordinate $\vec{r}$, the interaction will be:

$$H_{\text{eff}}(\vec{r}) = \int \rho_{\text{eff}}(\vec{r}\,')\, \phi_{\text{static}}(\vec{r}, \vec{r}\,')\, d\vec{r}\,', \tag{6.18}$$

where the static pion field of the proton is approximately:

$$\phi_{\text{static}}(\vec{r}, \vec{r}\,') = -\frac{\sqrt{2}f}{\sqrt{4\pi}} \tau_- \vec{\sigma} \cdot \vec{\nabla} \left(\frac{e^{-|\vec{r}-\vec{r}\,'|}}{|\vec{r}-\vec{r}\,'|} \right) \tag{6.19}$$

from (1.16). This interaction will cause the non-mesonic decay of the hyperon with ejection of nucleons, in complete analogy with ordinary internal conversion. With the approximation of one nucleon near the hyperon, the two nucleons in the final state (one nucleon being the "nucleus" in its ground state and the other being the ejected conversion "electron") will share the energy release of 177 Mev. This is a sufficiently large energy that the initial relative motion and the influence of other nucleons can be ignored, and the initial and final wave functions taken to be:

$$\psi_i = u_i \qquad \psi_f = u_f e^{i\vec{k}\cdot\vec{r}}$$

where u_i, u_f are the nucleon's spin wave functions (the hyperon's spin function is incorporated in ρ_{eff}), and the plane wave describes the relative motion in the final state. The wave number has the magnitude $k \simeq 2.9$ in units of μc.

The non-mesonic decay rate will be given by the square of the matrix element $\langle i|H_{\text{eff}}|f\rangle$, multiplied by the appropriate density of final states, summed and averaged over spins. The result is easily found to be:

$$\Gamma_n \simeq 8\pi\lambda^2 f^2 \frac{Mk^2}{(1+k^2)^2} k^{2\ell+1}. \tag{6.20}$$

The ratio of Γ_n to Γ_π of (6.16) will give the conversion coefficient for the decay mode $\Lambda^0 \to p + \pi^-$ per unit proton density at the position of the hyperon.[76] To obtain a conversion coefficient representing the ratio of hypernuclear non-mesonic decays to decays by emission of negative pions (the neutral pion mode is essentially unobservable), two corrections must be made: (a) Γ_n in (6.20) must be multiplied by the effective proton density ρ_p at the hyperon, (b) Γ_n must be multiplied by approximately $\frac{3}{2}$ to take into account the conversion by means of the static *neutral* pion field of the nucleons.[77] For a loosely bound Λ a rough estimate for ρ_p is given by:

$$\rho_p \simeq \left(\frac{Z}{\Omega}\right) \frac{\gamma R}{1+\gamma R}, \tag{6.21}$$

where Z is the atomic number, Ω is the nuclear volume, R is some sort of nuclear radius, and γ is the exponential decay constant of the bound state wave function ($\hbar^2\gamma^2 = 2M_\Lambda B$). The first factor ($Z/\Omega$) represents the proton density inside the nucleus, while the second is an estimate of the fraction of the time the hyperon spends inside the nucleus. The conversion coefficient can now be written:

$$Q \simeq \frac{3}{2}\frac{\Gamma_n}{\Gamma_\pi}\rho_p \simeq 12\pi f^2 \frac{Mk^2}{(1+k^2)^2}\rho_p\left(\frac{k}{q}\right)^{2\ell+1}. \tag{6.22}$$

In (6.22) we see the characteristic dependence on the multipole order of the transition. With $k/q \simeq 4.0$ this dependence is marked. Using $f^2 = 0.08$ we obtain the numerical estimate:

$$Q \simeq 7.4\,(16)^\ell \rho_p, \tag{6.23}$$

where ρ_p is expressed in units of the cube of the reciprocal of the pion Compton wavelength.

---

[76] The ratio of (6.20) to (6.16) is essentially the conversion coefficient obtained by M. Ruderman and R. Karplus, *Phys. Rev. 102*, 247 (1956).

[77] The branching ratios of 2 : 1 for decay into $p + \pi^-$ and $n + \pi^0$ show that the λ^2 value for π^0 emission is about $\frac{1}{2}$ of that for π^- emission. When the non-mesonic decay rate via the static neutral pion field is included $\Gamma_n\rho_p$ becomes $\Gamma_n\rho_p + \frac{1}{4}\Gamma_n(\rho_p + \rho_n)$, where the added factor of $\frac{1}{2}$ arises because τ_z replaces $\sqrt{2}\,\tau_-$ in the static field (6.19), and the proton and neutron densities both appear because they are equally effective for the interaction. With the approximation of $\rho_n \simeq \rho_p$, the revised estimate is $\frac{3}{2}\Gamma_n\rho_p$.

With the known binding energies of hypernuclei and data on proton densities from electron scattering, approximate effective proton densities can be calculated and conversion coefficients estimated. For hydrogen and helium hypernuclei these estimates are given in Table 7 for several ℓ values, along with the observed values.

Table 7. Estimated Conversion Coefficients (Non-mesonic decays/π^- decays) for Hydrogen and Helium Hypernuclei

	$\ell = 0$	$\ell = 1$	$\ell = 2$	Observed
H	0.3	5.3	85.	$0/6 \simeq 0$
He	0.9	14.0	222.	$16/7 \simeq 2.2$

The theoretical estimates are probably uncertain by a factor of 2 or 3, but the dependence on ℓ value is so rapid that the observations demand $\ell = 0$ or perhaps 1, corresponding to a Λ^0 spin of $\frac{1}{2}$, or perhaps $\frac{3}{2}$. With violation of parity conservation known to occur in decay processes, some mixture of $\ell = 0$ and $\ell = 1$ is completely consistent with spin $\frac{1}{2}$ for the Λ^0 hyperon. The data would imply that the $\ell = 0$ mode is dominant. The value of Λ spin from hypernuclear decays is in agreement with that found in the production and capture reactions (Section 6.2), and is useful because it does not depend on a spin assignment for the K-meson.

6.6. Neutral K-Mesons

In the Gell-Mann scheme there are two neutral K-mesons: K^0 and $\overline{K^0}$, with opposite strangeness. In strong interactions these particles will always be distinct since conservation of strangeness implies that they cannot be connected by virtual processes. Thus, in ordinary reactions involving the production of a single hyperon and K-meson, it will be the K^0 that is produced. In the decay processes, however, strangeness is not conserved. Consequently the K^0 and $\overline{K^0}$ can be connected together via virtual decay modes. This leads immediately to the prediction[78] that the decay of K^0 (or $\overline{K^0}$) should show two distinct lifetimes, in complete analogy with eigenfrequencies of coupled oscillators, etc. We define the decay mode with the shorter lifetime to be the K_1^0 "particle", and assume that it has the observed lifetime of $\sim 10^{-10}$ sec and dominant decay mode $K_1^0 \rightarrow \pi^+ + \pi^-$. The longer lifetime "particle" is called K_2^0. Its decay modes are not well known, but will not be the same as K_1^0.

We have the unusual situation that for the strong interactions the K^0 and $\overline{K^0}$ are the meaningful entities, while for weak decay interactions the K_1^0 and $\overline{K_2^0}$ are the empirically important quantities. The K^0 and $\overline{K^0}$ can

[78] M. Gell-Mann and A. Pais, *Phys. Rev.* *97*, 1387 (1955).

be expressed as linear combinations of K_1^0 and K_2^0. At the instant of production a K^0-meson will be represented by

$$K^0 = \frac{1}{\sqrt{2}} (K_1^0 + iK_2^0). \tag{6.24}$$

At some later time the wave function can be written

$$K^0 = \frac{1}{\sqrt{2}} (K_1^0 e^{-\lambda_1 t - i\omega_1 t} + iK_2^0 e^{-\lambda_2 t - i\omega_2 t}), \tag{6.25}$$

where $(\lambda_j + i\omega_j)$ is the complex decay constant (ω_j represents a mass shift) for the K_j^0 particle. If λ_2 is small compared to λ_1, only one half of the K^0-mesons produced in a target should decay with the 10^{-10} sec lifetime. Eisler et al.[50] have studied Λ^0 and K^0 production and decay in a propane bubble chamber. With the assumption of associated production, they find that 51 ± 8 percent of K^0's escape from the chamber before decaying, completely consistent with (6.25). From the dimensions of the chamber and the absence of "anomalous" θ^0 decays they can set a lower limit of $\sim 3 \times 10^{-8}$ sec for the lifetime of the K_2^0 decay mode. Other Brookhaven experiments[79] have established that the K_2^0 lifetime is less than 10^{-6} sec, and have given evidence for a predominance of three body modes ($\pi e \nu$, $\pi \mu \nu$, and $\pi\pi\pi$) in its decay. In spite of the present lack of detailed information on the K_2^0-meson, the existence of two distinct lifetimes in the decay of K^0-mesons is a striking confirmation of the Gell-Mann scheme. Even more bizarre manifestations of the mixing of K^0 and $\overline{K^0}$, involving regeneration of the short lived mode at a distance far from the point of production[80] or interference effects due to differences in the mass shifts,[81] have been proposed.

6.7. Decay Branching Ratios and the $\Delta T = \frac{1}{2}$ Rule

In the weak decay interactions the strangeness changes by one unit ($\Delta S = \pm 1$). Since charge and heavy particle number are absolutely conserved, (6.6) implies $\Delta T_z = \pm \frac{1}{2}$. It is natural to try to go one step further and postulate that $\Delta T = \pm \frac{1}{2}$, although there is no reason other than simplicity for this assumption. A certain number of predictions immediately follow.[82] For example, the Λ^0 decay branching ratio is predicted as $(\Lambda^0 \to p + \pi^-)/(\Lambda^0 \to n + \pi^0) = 2.0$, in excellent agreement

---

[79] Lande, Booth, Impeduglia, Lederman, and Chinowsky, *Phys. Rev. 103*, 1901 (1956); and Lande, Lederman, and Chinowsky, *Phys. Rev. 105*, 1925 (1957).

[80] A. Pais and O. Piccioni, *Phys. Rev. 100*, 1487 (1955); M. L. Good, *Phys. Rev. 105*, 1120 (1957).

[81] S. B. Treiman and R. G. Sachs, *Phys. Rev. 103*, 1545 (1956).

[82] The calculations can be most easily performed by assuming that isotopic spin is conserved in the decay, but that a "spurion" with $T = \frac{1}{2}$ and appropriate T_z is absorbed by the decaying particle. Then the well known vector addition coefficients can be employed to give branching ratios.

with the data (5.1). The $\Sigma^\pm$ decay branching ratios can similarly be shown to be consistent with this rule.

In the K-meson decays the approximate validity of the $\Delta T = \pm\frac{1}{2}$ rule gives a natural explanation of the great difference in lifetime between the neutral and charged θ modes ($\theta^0 \rightarrow \pi^+ + \pi^-$ or $\theta^0 \rightarrow \pi^0 + \pi^0$ and $\theta^\pm \rightarrow \pi^\pm + \pi^0$). The decay of the $\theta^\pm$ is essentially forbidden relative to the θ^0, being some 500 times slower. A priori this is difficult to understand. But we observe that the spin of the θ meson is very likely even. With two bosons in the final state the total wave function must be symmetric. Since the space part is symmetric, so also must be the isotopic spin part. With $\Delta T = \pm\frac{1}{2}$, the final state could have $T = 0$ or 1, but only $T = 0$ is symmetric. For θ^0 this state can occur since $T_z = 0$. But for $\theta^\pm$ the state cannot arise since $T_z = \pm 1$. Therefore $\theta^\pm$ decay is forbidden by the rule $\Delta T = \frac{1}{2}$. Electromagnetic processes will cause a breakdown of the rule, however, and it is plausible that the $\theta^\pm$ should have a lifetime several orders of magnitude greater than the θ^0 lifetime.

Application of the $\Delta T = \pm\frac{1}{2}$ rule to the K^0 branching ratios leads to the expectation that $K^0 \rightarrow \pi^0 + \pi^0$ should occur $\frac{1}{3}$ of the time, in contradiction with the experimental fraction of 0.12 ± 0.06.[50] In order to explain simultaneously this branching ratio and the large difference in lifetimes of the θ^0 and $\theta^\pm$ it is necessary to weaken the $\Delta T = \pm\frac{1}{2}$ rule to include $\Delta T = \pm\frac{3}{2}, \frac{5}{2}$.[83] The amplitudes of the admixtures of $|\Delta T| = \frac{3}{2}$ and $\frac{5}{2}$ are of the order of five to ten percent. We can still regard the $\Delta T = \pm\frac{1}{2}$ rule for decay processes as approximately valid, but we must apparently expect violations of the order of ten percent. Such comparatively large violations cannot be easily explained by means of electromagnetic effects. They indicate that the rule is probably not of a fundamental nature, but merely a good working hypothesis.

---

[83] M. Gell-Mann, *Nuovo Cimento*, *V*, 758 (1957).

CHAPTER 7

Theoretical Models for Strong Interactions

The empirical facts about hyperons and K-mesons have been discussed in the two preceding chapters in terms of a theoretical framework based on the idea of charge independence of interactions and the concept of strangeness. It was found that such a framework was very useful in correlating the data, and that the elementary properties of the particles and their interactions were consistent with the scheme. To go further and be able to predict or interpret detailed phenomena we need more than a framework; we need some sort of theory or model. In this chapter we will give a brief discussion of some of the attempts at theories, but will not go into great detail. The experiments are still comparatively crude, the problem is not clearly defined, and all theories proposed so far can be said to be just tentative beginnings. It may well be that attempts based on relativistic quantum field theory as we know it are bound to fail, and that an entirely new fundamental basis is required.

7.1. Goldhaber-Christy Model

The first theoretical model to be considered is a very simple one, proposed in outline by Goldhaber[84] several years ago to explain the existence of the hyperons, and recently extended to the interaction of K-mesons with nucleons by Christy.[85] It has some rather obvious flaws, as we shall see, but it provides a surprisingly good crude description of K-meson dynamics. The model is based on the assumption that the description of hyperons and K-mesons can be accomplished with the introduction of only *one* new particle, the K-meson.

The hyperons are assumed to be bound states of one or more K-mesons and nucleons. To obtain the correct strangeness for the hyperons it is obvious that the bound particles must be K^- and $\overline{K^0}$. Fortunately for the model, the observed interaction between K^- and nuclei is attractive (see Section 6.3), and the possibility of bound states can be visualized. The compound states of the hyperons with $S = -1$ can be listed as follows:

---

[84] M. Goldhaber, *Phys. Rev.* *101,* 433 (1956).

[85] R. F. Christy, footnote 49.

$$\begin{array}{lll} & \underline{\text{Hyperon}} & \underline{\text{Compound}} \\ T=1 & \left\{\begin{array}{l}\Sigma^+ \\ \Sigma^0 \\ \Sigma^-\end{array}\right. & \begin{array}{l}\overline{K^0}p \\ \frac{1}{\sqrt{2}}(\overline{K^0}n + K^-p) \\ K^-n\end{array} \\ T=0 & \Lambda^0 & \frac{1}{\sqrt{2}}(\overline{K^0}n - K^-p)\end{array} \tag{7.1}$$

The decay of the hyperons is envisioned as proceeding through the decay of the K-mesons and the strong interactions. For example,

$$\Sigma^+ = (\overline{K^0}p) \xrightarrow[\text{weak}]{} (\pi + \pi + p) \underset{\text{strong}}{\begin{array}{l}\nearrow \pi^+ + n \\ \searrow \pi^0 + p\end{array}} . \tag{7.2}$$

Production is imagined to follow the chain:

$$\begin{array}{ll}\text{photo} & \gamma + p \to (K^+ + K^- + p) \to K^+ + \Lambda^0 \text{ (or } \Sigma^0) \\ \text{mesonic} & \pi^- + p \to (\pi^- + K^+ + K^- + p) \to K^+ + \Sigma^- .\end{array} \tag{7.3}$$

The masses of the hyperons can be correlated in a simple way if we assume that there is some sort of potential acting between K-mesons and nucleons. In first approximation the binding energy is linearly proportional to the number of bosons in the potential well. Empirically we know that the potential (or the binding) depends on $(\boldsymbol{\tau}_1 \cdot \boldsymbol{\tau}_2)$ since the Σ and Λ have different masses. Thus the difference between the hyperon masses and that of a nucleon can be written in the form:

$$M_H - M_N = -SA + (\boldsymbol{\tau}_N \cdot \sum_i \boldsymbol{\tau}_{K_i})B , \tag{7.4}$$

where S is the strangeness (number of K-mesons bound), and A and B are empirical constants. From the Σ, Λ masses (treating the $\Sigma^{+,0,-}$ as degenerate) we find $A \simeq 234$ Mev, $B \simeq 77$ Mev.

For the Ξ particle $S = -2$, i.e. two K-mesons bound around a nucleon. Two K's can combine to give $\boldsymbol{T}_K = \sum_i \boldsymbol{\tau}_{K_i}$ with eigenvalues 0 and 1. For $T_K = 0$, the total isotopic spin of the hyperon will be $T = \frac{1}{2}$ and $\boldsymbol{\tau}_N \cdot \boldsymbol{T}_K = 0$. For $T_K = 1$, there are two values of T: $T = \frac{1}{2}$, with $\boldsymbol{\tau}_N \cdot \boldsymbol{T}_K = -1$, and $T = \frac{3}{2}$, with $\boldsymbol{\tau}_N \cdot \boldsymbol{T}_K = \frac{1}{2}$. Consequently the lowest energy eigenvalue has $T_K = 1$, $T = \frac{1}{2}$ (an isotopic spin doublet, as "observed"), and a mass excess $\Delta M = 2(234) - 1(77) = 391$ Mev. This can be compared with the observed mass excess of the Ξ^- of 382 Mev. This is good enough agreement, but there are difficulties in that there should be higher states at $M_\Xi + 77$ Mev and $M_\Xi + 115$ Mev. One can argue that the $M_\Xi + 77$ Mev doublet would decay rapidly by gamma emission to the Ξ, but the quartet at $M_\Xi + 115$ Mev should be as stable as the other hyperons (It would, incidentally, have a member with a double negative charge). The model gives no explanation of these discrepancies.

Turning now to K-meson scattering by protons, we recall that the K^+ scattering up to $\sim$100 Mev was roughly isotropic in the center of mass, and essentially constant in energy at $\sim$15 mb. These data imply s-wave scattering by a short-range potential. The K^+ and p can only form a $T = 1$ state. Consequently, Christy assumes that the $K^+ - p$ interaction is equal to the interaction which gives the bound Σ particle, but of the opposite sign (S is opposite, and empirically the K^+-nucleus interaction is repulsive). Taking a square well, the binding energy of the Σ (241 Mev) and the 15 $\pm$ 5 mb total cross section at zero energy are sufficient to determine the depth and radius of the well ($V_0 = 1000 \pm 200$ Mev, $R = 0.6 \pm 0.1 \times 10^{-13}$ cm). The resulting cross section is essentially constant in energy, having fallen from 15 mb to 13.9 mb at 100 Mev. For $K^+ + n$ scattering, Christy argues that the ordinary scattering will be essentially the same as for $K^+ + p$, while the exchange scattering ($K^+ + n \rightarrow K^0 + p$) will have a very small cross section. This results essentially from the weak dependence of the potential on $\boldsymbol{\tau}_N \cdot \boldsymbol{\tau}_K$.

For negative K scattering on nucleons the cross section is considerably larger, of the order of 70 mb at zero energy. In the Goldhaber-Christy model this is a consequence of the attractive nature of the K^--nucleon interaction as opposed to the repulsion in K^+-nucleon scattering. The same square well parameters as for K^+-nucleon scattering (but as an attraction) give a scattering cross section of 70 mb at zero energy, in agreement with observation, but the calculated cross section falls off with energy much more slowly than the data in Table 6. The description of scattering in terms of a simple real potential precludes a treatment of the K^--reaction cross sections.

The model can be pushed further and used to discuss with some success the lifetimes of the hyperons, the photo-production of K-mesons and hyperons, the absorption of K^- by nucleons, and K production by pions, etc. It is a useful device for orientation purposes, but cannot really be taken too seriously. Aside from its lack of basis from a fundamental point of view, there are obvious inadequacies in treating by static potentials interactions which give binding energies comparable to the mass of the constituent particles. What success the model has in correlating scattering data is probably shared by any short range interaction in S-states.

7.2. Form of the K-Meson—Baryon Interaction

In analogy with the electromagnetic field and the pion field it is natural to attempt to write down an interaction between the K-meson (boson) field and the baryon (fermion) fields which is linear in the K-meson field and satisfies all the proper invariance requirements. As we saw in Section

1.2 for the pions, the specific form of the interaction depends on such properties as the intrinsic parity of the boson (and the fermions). Since the pion is pseudoscalar, the appropriate relativistic coupling to the nucleons involved the pseudoscalar coupling operator ($i\gamma_5$). We must therefore discuss the question of the intrinsic parities of the K-mesons and hyperons. In the strong interaction production process the K-mesons and/or hyperons are always produced in pairs. Consequently their individual parities relative to the nucleon cannot be defined. But once the parity of one particle is specified, the parity of all the others relative to it can in principle be determined. It is conventional to define the parity of the Λ hyperon relative to the nucleon as positive, and express the parities of the other hyperons and K-mesons in terms of it.

The emission and absorption of K-mesons by baryons can be symbolized by the equations:

$$\begin{aligned} N &\leftrightarrow \Lambda + K \\ N &\leftrightarrow \Sigma + K \\ \Lambda &\leftrightarrow \Xi + K \\ \Sigma &\leftrightarrow \Xi + K\,. \end{aligned} \tag{7.5}$$

These relations, and those obtained from them by transferring the symbols from one side of the equation to the other (and at the same time putting a bar on the symbol to mean antiparticle), are all the possible emission and absorption acts allowed by the strangeness concept. We can express these basic processes in terms of an interaction Lagrangian density:

$$\mathscr{L}_K = \sqrt{4\pi}[g_{\Lambda K}(\bar{N}\Lambda)K + g_{\Sigma K}(\bar{N}\Sigma)K + h_{\Lambda K}(\bar{\Lambda}\Xi)K + h_{\Sigma K}(\bar{\Sigma}\Xi)K] + \text{h.c.}\,, \tag{7.6}$$

where h.c. means Hermitian conjugate, and the particle symbol is the field operator that destroys the particle (and creates the antiparticle). The constants g and h are coupling constants. The form of (7.6) is only a symbolic skeleton with details such as the interaction operators and the charge dependence suppressed.

Until the parities of the K-meson and the Σ and Ξ hyperons are specified, the actual interaction operators to be inserted cannot be determined. From the assumption of charge independence of the interactions, however, the particular charge dependence can be written down easily. For each of the terms in (7.6) the problem is merely one of coupling the isotopic spins of the two particles being destroyed to add up to the isotopic spin of the particle being created. For example, the first term can be written out:

$$(\bar{N}\Lambda)K = [(\bar{p}\Lambda)K^+ + (\bar{n}\Lambda)K^0] + \text{h.c.}\,, \tag{7.7}$$

where the symbol K^+ corresponds to absorption of K^+ or emission of

$\overline{K^+} = K^-$, and similarly for the symbol K^0. For the nucleon-sigma coupling the result is:

$$(\bar{N}\Sigma)K = [\sqrt{2}(\bar{p}\Sigma^+)K^0 + (\bar{p}\Sigma^0)K^+ - (\bar{n}\Sigma^0)K^0 + \sqrt{2}(\bar{n}\Sigma^-)K^+] + \text{h.c.} \tag{7.8}$$

The couplings of the Ξ hyperon to Λ and Σ are of the same form with ($n \to \Xi^-$, $p \to \Xi^0$, $K^0 \to \overline{K^+}$, $K^+ \to \overline{K^0}$).

In an attempt to learn something about the parity of the K-meson and the type of interaction in (7.6), Wentzel[86] examined the question of the binding of Λ hyperons in hypernuclei, assuming that the K-baryon coupling was responsible. The first order approximation to the Λ hyperon-nucleon force is shown diagrammatically in Fig. 13. It corresponds to the

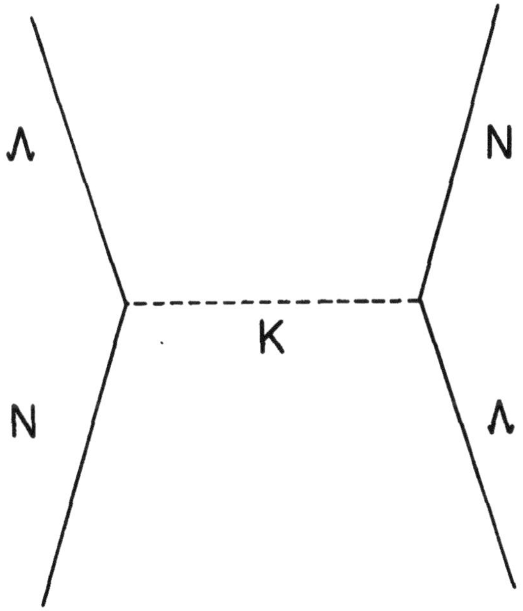

Fig. 13. Feynman diagram for the lowest order contribution to the Λ hyperon-nucleon force.

exchange of a single virtual K-meson, according to (7.7). In the static limit the potential involves the spatial dependence $(e^{-\mu r}/r)$, where μ is essentially the reciprocal Compton wavelength of the K-meson. The exchange of a single K-meson implies that the roles of the two baryons are interchanged, as Fig. 13 shows. This means that the potential will involve both spin and space exchange operators, $\frac{1}{2}(1 + \vec{\sigma_1} \cdot \vec{\sigma_2})$ and P_{12}, familiar from ordinary nuclear physics, as well as any spin dependence

[86] G. Wentzel, *Phys. Rev.* *101*, 835 (1956).

arising from the specific interaction form. If the Λ hyperon has spin ($\frac{1}{2}+$) and the K-meson ($0\pm$), the two-body static interaction is given in Table 8 for the various spin and parity states of the two baryons and the two choices of K-meson parity. With scalar K-mesons, the mesons are emitted as s-waves and there is no spin flip. The space and spin exchange is given entirely by $\frac{1}{2}(1 + \vec{\sigma_1} \cdot \vec{\sigma_2})P_{12}$, and for the S states gives an attractive potential in the singlet spin state, but repulsive in the triplet.

Table 8. Lowest Order Central Force Between Λ Hyperon and Nucleon

$V(r) = g^2 s \dfrac{e^{-\mu r}}{r}$, values of s are tabulated.

Spin-Parity of K-meson	Coupling Type	Even Parity		Odd Parity	
		Singlet	Triplet	Singlet	Triplet
0+	K	−1	+1	+1	−1
0−	$\vec{\sigma} \cdot \vec{\nabla} K$	−3	−1	+3	+1

For pseudoscalar K-mesons, p-wave emission with spin flip occurs. The additional spin dependence makes the potential attractive for both singlet and triplet S states, but more attractive in the singlet.

The analysis of hypernuclear binding (see Section 6.4) for the highest nuclei is consistent with the pseudoscalar K-meson, but not with the scalar. The coupling constant has to be of the order of the pion-nucleon coupling or even larger. Unfortunately, there are several complications. First of all, from our experience with the nucleon-nucleon force problem it is safe to say that lowest order perturbation theory cannot be trusted to give more than a very qualitative description of the interaction, especially if the coupling constant is large. Secondly, the spatial exchange character implied by the single K-meson exchange can probably be ruled out from an examination of the Λ hyperon binding in nuclei such as Li^{7*} where p-state nucleons contribute to the force. Finally, as will be discussed in the next section, it seems likely that the Λ binding in hypernuclei can be explained in terms of pion rather than K-meson interactions. As a result, the form and strength of the K-meson-baryon interactions, as well as the parity of the K-meson, cannot be safely determined by study of the Λ-nucleon forces.

Another approach at the K-meson-baryon coupling is through examination of K-meson-nucleon scattering. The problem is considerably more difficult than the pion-nucleon problem: the large mass of the K-meson relative to the nucleon ($M_K = 0.525\ M_N$) makes recoil terms very important; the strong pion-baryon interaction (see the next Section) complicates matters; nucleon pair contributions are presumably more significant.

Some preliminary investigations have been performed[87] in the (no-recoil, no pairs, no pion-baryon interaction) approximation, with various assumptions about the parities of the K relative to the Λ and Σ. For pseudoscalar K-mesons the coupling is taken to be pseudovector $(\vec{\sigma} \cdot \vec{\nabla} K)$. With suitable choice of the coupling strengths ($g^2_{\Lambda K} \simeq g^2_{\Sigma K} \simeq 0.3$), the calculated *cross sections* for $K^+ - p$ scattering are in rough agreement with experiment at energies of the order of 100 Mev for both scalar and pseudoscalar K-mesons, with a considerably larger contribution from the $T = 1$ state than the $T = 0$ state in accord with the indications of Section 6.3. An apparent difficulty is that the sign of the dominant phase shift implies an *attraction* between K^+-meson and nucleon, contrary to the evidence from Coulomb interference effects. Rough estimates on the K^--nucleon interaction give rise to similar difficulties, the dominant phase shifts implying a repulsion rather than the attraction indicated experimentally.

It is evident that the K-meson-baryon interaction is a complicated three field (K-meson-pion-fermion) problem, and that much more trustworthy means of calculation must be developed before meaningful comparisons with experiment can be made.

7.3. Form of the Pion-Baryon Interaction and Hypernuclear Binding

The fundamental interaction of nucleons and pions is customarily written in terms of pseudoscalar coupling of the pion field to the nucleons (1.3), although the static model was used as the basis of our discussion in Part I. It is reasonable to expect that all the baryons should interact with the pion field in the same basic way, and we can ask about the form of the coupling if charge independence is again assumed. The elementary processes of pion emission and absorption by the hyperons are:

$$\begin{aligned} \Lambda &\leftrightarrow \Sigma + \pi \\ \Sigma &\leftrightarrow \Sigma + \pi \\ \Xi &\leftrightarrow \Xi + \pi, \end{aligned} \tag{7.9}$$

the process $\Lambda \leftrightarrow \Lambda + \pi$ being forbidden by conservation of isotopic spin. With the notation of vectors in charge space the charge-independent forms for the interaction Lagrangian density can be written down directly:

$$\mathscr{L}_\pi = \sqrt{4\pi}[g_\Lambda(\bar{\mathbf{\Sigma}}\Lambda) \cdot \boldsymbol{\pi} + ig_\Sigma(\bar{\mathbf{\Sigma}} \times \mathbf{\Sigma}) \cdot \boldsymbol{\pi} + g_\Xi(\bar{\Xi}\boldsymbol{\tau}\Xi) \cdot \boldsymbol{\pi}] + \text{h.c.}, \tag{7.10}$$

where the convention is similar to that employed in (7.7) and (7.8). The specific interaction operators have been omitted. If the Σ and Λ have the same parity, the operator $(i\gamma_5)$ is to be inserted between the fermion field

[87] D. Amati and B. Vitale, *Nuovo Cimento, V*, 1533 (1957), and C. Ceolin and L. Taffara, *Nuovo Cimento, VI*, 425 (1957).

operators in each term, while if they have opposite parity the first term will have the operator (1) in it. For definiteness we write out the separate interactions in (7.10) explicitly with the individual particle operators:

$$\mathscr{L}_{\pi\Lambda} = \sqrt{4\pi}\, g_\Lambda[(\bar{\Sigma}^+\Lambda)\pi^+ + (\bar{\Sigma}^0\Lambda)\pi^0 + (\bar{\Sigma}^-\Lambda)\pi^-] + \text{h.c.}, \qquad (7.11)$$

$$\mathscr{L}_{\pi\Sigma} = \sqrt{4\pi}\, g_\Sigma[(\bar{\Sigma}^0\Sigma^- - \bar{\Sigma}^+\Sigma^0)\pi^+ + (\bar{\Sigma}^+\Sigma^+ - \bar{\Sigma}^-\Sigma^-)\pi^0 + (\bar{\Sigma}^-\Sigma^0 - \bar{\Sigma}^0\Sigma^+)\pi^-] + \text{h.c.}, \qquad (7.12)$$

$$\mathscr{L}_{\pi\Xi} = \sqrt{4\pi}\, g_\Xi[\sqrt{2}(\bar{\Xi}^0\Xi^-)\pi^+ + (\bar{\Xi}^0\Xi^0 - \bar{\Xi}^-\Xi^-)\pi^0 + \sqrt{2}(\bar{\Xi}^-\Xi^0)\pi^-] + \text{h.c.} \qquad (7.13)$$

This last form is completely equivalent to the pion-nucleon interaction with the replacements ($\Xi^- \rightarrow n$, $\Xi^0 \rightarrow p$).

The contribution of the pion-baryon interaction to the binding of Λ hyperons in nuclei has been considered by Lichtenberg and Ross[88] and Dallaporta and Ferrari.[89] The lowest order pion exchange process involves *two* pions, as indicated in Fig. 14, and consequently gives a short-range

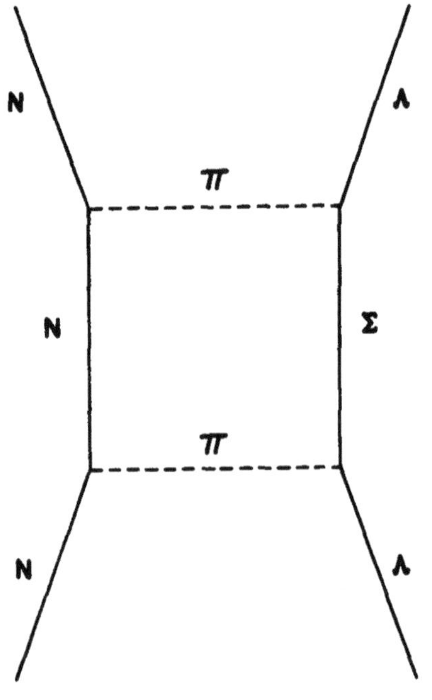

Fig. 14. Feynman diagram for the lowest order pion contribution to the Λ hyperon-nucleon force.

---

[88] D. B. Lichtenberg and M. Ross, *Phys. Rev. 103*, 1131 (1956), and *Phys. Rev. 107* 1714 (1957).

[89] N. Dallaporta and F. Ferrari, *Nuovo Cimento V*, 111 (1957).

($\sim\hbar/2m_\pi c$) ordinary potential akin to the fourth order contributions to the nucleon-nucleon force. The calculations of the potential were based on the static model, using a Hamiltonian of the form (1.18) with the appropriate isotopic spin dependence. With a suitable hard core at $\sim 0.4 \times 10^{-13}$ cm we can expect the potential to give a semi-quantitative description of the Λ hyperon-nucleon interaction.

The question of the parity of the Σ relative to the Λ arises, and two cases must be considered just as in Section 7.2. If the Λ and Σ have opposite parity, the pions emitted or absorbed by the Λ or Σ are *s*-wave pions. This leads to forces which are repulsive and independent of spin. If the Λ and Σ have the same parity, the *p*-wave emission and absorption of pions with spin-flip leads to attractive, spin-dependent forces with the singlet force almost twice as attractive as the triplet force (in terms of volume integrals).

The potential found for the case of the same parity for Λ and Σ are quite consistent with the binding of Λ hyperons in hypernuclei, provided the coupling constant g_Λ in (7.11) is very nearly equal to the pion-nucleon coupling constant g_N ($g_N^2 = \left(\frac{2Mf}{m_\pi}\right)^2 \simeq 15$). The theory implies that the singlet spin state between the nucleon and lambda is favored. With the values of $U_{\rm av} \simeq 220.$ Mev $\times 10^{-39}$ cm^3 and $U_{f{\rm av}} \simeq 350.$ Mev $\times 10^{-39}$ cm^3 from Section 6.4, we find $U_s \simeq 2U_t$ in reasonable accord with the calculations.

It seems reasonable and attractive to assume that:

(a) The parity of the Σ is the same as that of the Λ.

(b) All the baryons are coupled to the pions with approximately the same strength.

(c) The binding of Λ hyperons in nuclei is due predominantly to two pion exchange forces, rather than forces due to the exchange of *K*-mesons.

7.4. The Symmetric Pion Coupling Model

A possible model of strong couplings has been proposed independently by Gell-Mann[90] and Schwinger.[91] The concepts of ordinary field theory are employed, although the symmetry properties of the model are more general. The fundamental assumption is that the various interactions of physics can be classified according to their strengths, and that the strongest interactions have the highest degree of symmetry in their coupling of the particles. The strong interactions, as described in Section 6.1, are the coupling of the nucleons and hyperons to the pions and *K*-mesons. These interactions are believed to be charge independent, corresponding to invariance under rotations in isotopic spin space. The weaker electromagnetic interaction causes a breakdown of this symmetry by singling

[90] M. Gell-Mann, *Phys. Rev.* **106**, 1296 (1957).
[91] J. Schwinger, *Annals of Physics* **2**, 407 (1957).

out a preferred direction in isotopic spin space. The still weaker decay interactions violate further symmetries, such as space inversion symmetry as is discussed in Part III.

From the evidence of the preceding sections it is suggestive to argue that the pion-baryon interaction is the strongest ($g^2 \sim 15$) and the K-meson-baryon coupling somewhat weaker ($g^2 \sim 1$). If the pion-baryon coupling is assumed to have the highest possible symmetry, all the baryons must be coupled to pions in the same way. Then, in the absence of all weaker interactions (K-meson, electromagnetic, etc.), the nucleons and hyperons would all have the same mass and form an eight-fold degenerate supermultiplet.[92] It is straightforward to show that the requirement of identical couplings to the pion field is

$$g_\Lambda = g_\Sigma = g$$

and (7.14)

$$g_N^2 = g^2 = g_\Xi^2,$$

where g_Λ, g_Σ, and g_Ξ are the coupling constants in (7.11), (7.12), and (7.13), while g_N is the pion-nucleon coupling constant. The signs of g_N and g_Ξ relative to g are left open at this stage. They are relevant for such things as the Σ-nucleon interaction potential, but do not matter for the mass degeneracy.

The observed mass splitting of the baryons is attributed to the lower degree of symmetry of the coupling of the K-mesons to the baryons. The interaction can be assumed to be of the form of (7.6), with different coupling constants $g_{\Lambda K}$, $g_{\Sigma K}$, $h_{\Lambda K}$, and $h_{\Sigma K}$ in general. The specific interaction operator chosen will depend on the parity of the K-meson. The lowest order mass shifts caused by this K-meson-baryon interaction are such that the center of gravity of the baryon masses remains unaltered, each charge state of a multiplet being given unit weight. Thus we expect the following relation to hold approximately:

$$\tfrac{1}{2}(M_N + M_\Xi) \simeq \tfrac{1}{4}(M_\Lambda + 3M_\Sigma). \qquad (7.15)$$

Experimentally the left-hand side lies 191 Mev above the nucleon mass, while the right-hand side is 234 Mev above. The difference of 43 Mev in an overall splitting of 382 Mev is not unreasonable in view of the fact that (7.15) holds only in lowest order and the K-meson-baryon coupling is presumably not weak.

The Gell-Mann—Schwinger model offers the possibility that our experience with the more familiar pion-nucleon problem can be exploited to handle the strong pion-baryon interaction more or less exactly, with experimental data from pion physics used where possible, while the weaker K-meson-baryon interactions can be treated by perturbation techniques.

---

[92] The parity of the Σ is, of course, assumed to be the same as that of the Λ.

Even if such an attempt fails and a radically different approach is demanded, the underlying symmetry properties of the model may remain intact.

Tiomno[93] has considered a related model possessing a high degree of symmetry for all the strong interactions. He finds that the observed mass splittings can be obtained with interactions of the strength of electromagnetic, rather than K-meson, couplings.

7.5. Photoproduction of K-Mesons from Hydrogen and the Parity of the K-Meson

The development of a detailed description of K-meson dynamics must await more information on the fundamental properties of the particles involved. In particular, the evidence presented so far has cast little light on the parity of the K-meson. One weak argument *against* the K-meson being *scalar* might be that the calculated K-meson-nucleon interaction in the dominant $T = 1$ state is attractive, while experimentally it seems repulsive (see Section 7.2). Of course, the calculations for a pseudoscalar K-meson with pseudovector coupling also gave an attraction. But the argument is that perhaps we can trust a scalar coupling calculation more than a pseudovector coupling calculation. Whatever the merits of this line of reasoning, preliminary evidence on the photo-production of K-mesons from hydrogen seem to favor a scalar K-meson.

The photoproduction process studied is

$$\gamma + p \rightarrow \Lambda^0 + K^+. \tag{7.16}$$

No detailed theory of photoproduction, such as described in Chapter 4 for pions, is available. Only lowest order perturbation calculations have been made.[94] Four cases are considered: Scalar and pseudoscalar K-mesons, without and with anomalous magnetic moments for the baryons. Without the anomalous magnetic moment contributions, the two possibilities for the parity of the K-meson yield qualitatively different photoproduction cross sections. For scalar K-mesons the dominant feature is the direct photoelectric production via the K-meson current (corresponding to the pion current term in (4.25)). This gives predominantly a "retarded" $\sin^2 \theta$ distribution:

$$\left(\frac{d\sigma}{d\Omega}\right)_{\text{scalar}} \sim \frac{\sin^2 \theta}{(1 - \beta \cos \theta)^2}. \tag{7.17}$$

The pseudoscalar case without anomalous moments is superficially like the pion photoproduction at low energies, yielding a more or less isotropic differential cross section at energies less than 100 Mev above threshold.

---

[93] J. Tiomno, footnote 49; see also his earlier paper in *Nuovo Cimento, VI*, 69 (1957).

[94] M. Kawaguchi and M. J. Moravcsik, *Phys. Rev. 107*, 563 (1957); A. Fujii and R. E. Marshak, *Phys. Rev. 107*, 570 (1957).

Preliminary experimental data on the angular distribution are available[95] at 1000 Mev photon energy. This is about 90 Mev above threshold. The results are consistent with the retarded $\sin^2\theta$ distribution, but are not definitive because the cross section was measured only at backward angles (70°, 90°, and 155°).

While the experiments tend to favor a scalar K-meson, the situation is complicated by the fact that the theoretical calculations including anomalous magnetic moments for the baryons are very sensitive to the specific choice of the moment for the Λ^0, at least for the odd parity case. The angular distribution can change character markedly with slight changes in numerical values of $\mu(\Lambda^0)$. When there are more data, especially at small angles, and more reliable calculations, perhaps the parity of the K-meson can be established from photoproduction.

---

[95] P. L. Donoho and R. L. Walker, *Phys. Rev. 107,* 1198 (1957); Silverman, Wilson, and Woodward, *Phys. Rev. 108,* 501 (1957).

PART III

DECAY PROCESSES

CHAPTER 8

Symmetry Principles and their Observational Tests

In the interpretation of the multitude of different decay modes of K-mesons discussed in Chapter 5 we saw that the $\tau - \theta$ puzzle (the θ "must" have parity $(-)^J$, the τ "cannot" have parity $(-)^J$) could be resolved easily, provided one allowed the idea of parity conservation to be discarded in decay processes. In a now famous paper, Lee and Yang[51] considered the evidence for or against parity conservation in decay processes and found no evidence one way or the other. They then suggested a number of experimental tests in the decay of hyperons and K-mesons as well as beta decay. Experiments[96] soon showed in a most dramatic way that space inversion invariance does not hold in beta decay and pi and mu meson decay. Since then, parity non-conservation has been observed to occur in hyperon decay and in the $K_{\mu 2}$ mode of K-meson decay. There is therefore good evidence for decisive lack of parity conservation in all weak decay processes.

Before describing the experiments and theory we will give some general considerations on the validity of the concept of parity for strong interactions and the observational quantities necessary to detect a parity violation.

8.1. Parity Conservation in Strong Interactions

We recall that thirty years ago Wigner introduced the formal idea of the parity of a quantum state as a generalization of Laporte's rule of atomic spectra (odd terms combine with even, and vice versa). In the interval, the concept has permeated our thinking very thoroughly and has been experimentally verified in many ways. Parity conservation forms the basis of the standard arguments in the text books[97] for the existence of only even static electric multipole moments and odd magnetic ones. One experiment quoted in this regard is that of Smith, Purcell, and Ramsey[98] who looked for an electric dipole moment of the neutron and found that it was less than 10^{-19} e.cm, whereas on simple arguments of size, $\sim 10^{-13}$ e.cm would be expected. It is now known that the electric dipole moment vanishes if either time reversal or space inversion invariance

[96] Experimental references will be given at the appropriate places in the detailed discussion below.

[97] See, for example, J. M. Blatt and V. F. Weisskopf, *Theoretical Nuclear Physics*, Wiley, New York, 1952, pp. 30 and 35.

[98] J. H. Smith, E. M. Purcell, N. F. Ramsey, *Phys. Rev. 108*, 120 (1957).

holds, and this example must be interpreted in that way. Perhaps it is better to quote a recent experiment by Tanner[99] which showed that in nuclear reactions parity is conserved to about 1 in 10^7 in intensity. He studied the yield of the highest energy group of alpha particles in the reaction

$$p + F^{19} \rightarrow O^{16} + \alpha \tag{8.1}$$

in the neighborhood of a known 1^+ state in the compound nucleus Ne^{20}. Since the spins of O^{16} and the alpha particle are both 0^+, they can come only from compound states in Ne^{20} with spin and parities $(0^+, 1^-, 2^+, \cdots)$, provided parity is conserved in the reaction. No evidence was found for alpha particles from the resonance in question, implying parity conservation within the experimental limits of detection.

At the present time it appears that parity conservation holds for strong interactions to a high degree of approximation. There will be, of course, very slight departures due to virtual decay processes, but these are only of the order of 10^{-24}, well beyond present limits of observation.

8.2. General Arguments on Invariance

It is of interest to consider what observational quantities we expect to see in a decay process, assuming various types of invariance requirements. A general decay process will have a number of particles emerging in various directions, and with different energies, and perhaps spins. One can conceive of many different observational quantities, but the invariance of the system under various transformations will restrict the type of quantity that can occur. We consider first rotational invariance.

(a) Rotational Invariance

Since the decaying system or state is assumed not to know or care which way we choose our coordinate axes for purposes of measurement, it is clear that the distribution function for the decay must be rotationally invariant, (i.e. a scalar under *rotations*). In ordinary beta decay, for example, the only vectors present are the momentum vectors of the electron and neutrino $(\vec{p}, \vec{q})$, and the only rotationally invariant combinations are $\vec{p} \cdot \vec{p} = p^2$, $\vec{q} \cdot \vec{q} = q^2$, and $\vec{p} \cdot \vec{q}$. Therefore, if only the momenta of electron and neutrino are observed, the distribution function will be of the form:

$$\sum_{j=0}^{N} F_j(\vec{p} \cdot \vec{q})^j, \tag{8.2}$$

where $F_j = F_j(p^2, q^2, \zeta)$ are functions of scalar quantities, and N depends on the degree of forbiddenness of the decay ($N = 1$ for allowed beta decay).

[99] N. Tanner, *Phys. Rev. 107*, 1203 (1957). See also the recent work on reactions in light nuclei by D. H. Wilkinson, *Phys. Rev. 109*, 1603, 1610, 1614 (1958).

(b) Space Inversion

Under space inversion we must take more care and distinguish between scalar and pseudoscalar quantities, vector and pseudovector (or axial vector) quantities. Examples are:

Scalar—pure number, scalar product of two vectors.

Vector—coordinate, linear momentum, electric field.

Pseudovector—angular momentum, vector product of two vectors, magnetic field.

Pseudoscalar—scalar product of a vector and a pseudovector, triple scalar product of three vectors.

If the decay process is invariant under space inversion, the distribution function must be made up of true scalars, *not* pseudoscalars. For example, in the study of the angular correlation of gamma radiation from oriented nuclei, the relevant quantities are the nuclear angular momentum vector $\vec{J}$ and the photon momentum $\vec{k}$. The scalar product $\vec{J} \cdot \vec{k}$ is a pseudoscalar. Consequently it must appear always to an even power, and the distribution is of the form:

$$\sum_{n=0}^{N} A_n (\vec{J} \cdot \vec{k})^{2n}. \tag{8.3}$$

We note that the distribution is symmetric above and below the plane defined by $\vec{J}$ and depends on the nuclear alignment but not the nuclear orientation (the sign of $\vec{J}$ is not significant).

If space inversion invariance does not hold, the distribution will not in general be invariant under inversion and may contain pseudoscalar quantities. For the beta decay of oriented nuclei, the term $(\vec{J} \cdot \vec{p})$ can occur. As pointed out by Lee and Yang,[51] for allowed beta transitions the distribution function is of the form:

$$F_1 + F_2(\vec{J} \cdot \vec{p}). \tag{8.4}$$

This distribution is no longer symmetric above and below the plane defined by $\vec{J}$. Another possibility is the pseudoscalar $(\vec{\sigma} \cdot \vec{p})$ where $\vec{\sigma}$ is the spin of the beta particle. The presence of this term in the distribution function is equivalent to saying the beta particles are longitudinally polarized. Detection of the directional asymmetry implied by (8.4) and the longitudinal polarization of the electrons have provided proof of the violation of inversion symmetry in beta decay.

8.3. Time Reversal Invariance

Another invariance property is that of time reversal. It is generally assumed that an isolated quantum-mechanical system cannot "sense" the direction of time, i.e. given a solution of the equations of motion, the quantum state obtained by changing $t \to -t$ is an equally acceptable solution of the same equations of motion.[100]

The important point is that under time reversal the sense of motion is reversed but the position is not changed, so that coordinate vectors are unaffected but momentum vectors such as $\vec{p}$, $\vec{J}$, $\vec{\sigma}$ all change sign under time reversal. In analogy with the arguments given above, we see that if decay processes are invariant under time reversal, only scalar (or pseudoscalar) products involving an *even* number of such vectors (or pseudovectors) can appear in the distribution function. For example, terms like $(\vec{J} \cdot \vec{\sigma})$, $(\vec{J} \cdot \vec{p})$, $(\vec{p} \cdot \vec{q})$, and $(\vec{J} \cdot \vec{p})(\vec{\sigma} \cdot \vec{p})$ can occur. If time reversal invariance is violated, then we can have typical terms like $\vec{J} \cdot (\vec{p} \times \vec{q})$ in the beta decay of oriented nuclei, or $\vec{J} \cdot (\vec{p} \times \vec{k})(\vec{J} \cdot \vec{k})$ in the beta-gamma angular correlation from aligned nuclei.[101]

A word of caution must be inserted here. The arguments just given about "time reversal testing" terms in the distribution function need modification if there is final state interaction, such as the Coulomb interaction for beta particles in beta decay. Instead of plane waves completely specified by the momentum vector $\vec{p}$, one must consider the exact solutions which consist asymptotically of a plane wave plus an *incoming* spherical wave. The operation of time reversal turns such a solution into one which asymptotically is a plane wave of the opposite momentum plus an *outgoing* spherical wave, and the simple arguments about vectors are not sufficient. The effect is that terms in the distribution function (which, in the absence of final state interactions, "test" time reversal invariance) get corrections which would be present even if time reversal invariance holds, and terms which did not "test" time reversal invariance get corrections which are present only if time reversal invariance fails to hold. For beta decay the corrections are of order $(Z/137)$ and are not very important in general,[102] at least for allowed transitions.

Time reversal invariance can manifest itself in other ways. In electromagnetic decay processes the assumption of time reversal invariance puts restrictions on the phases of matrix elements. In mixed multipole transitions the phase difference between the interfering multipole amplitudes must be 0° or 180°. Similarly, on the basis of unitarity of the S matrix and time reversal invariance in pion photoproduction from nucleons, the phases

---

[100] The actual formal operation is a bit more complicated than merely $t \to -t$. For a particle satisfying the Dirac equation, for example, the time reversed state is defined by $\psi_{TR}(\vec{x},t) = i\sigma_2\psi^*(\vec{x}, -t) = \gamma_3\gamma_1\psi^*(\vec{x}, -t)$ in the usual (Pauli) representation.

[101] At first glance a term like $\vec{\sigma} \cdot \vec{p}_i \times \vec{p}_f$, appearing in the polarization of protons or neutrons in a nuclear reaction, seems to violate time reversal. It does not, however, because under time reversal the roles of the initial and final states are interchanged $(\vec{p}_i \to -\vec{p}_f, \vec{p}_f \to -\vec{p}_i)$.

[102] Jackson, Treiman, and Wyld, *Nuclear Physics 4*, 206 (1957).

of the amplitudes leading to states of definite energy, spin, parity, and isotopic spin, are equal to the corresponding phase shifts for pion-nucleon scattering. As was already mentioned, time reversal invariance is sufficient to exclude static electric dipole moments for the neutron and other elementary particles, even though parity is not conserved.[103]

8.4. PCT Theorem and Charge Conjugation Invariance

In conventional quantum field theory one usually demands the theory to be not only space inversion and time reversal invariant, but also charge conjugation invariant. Charge conjugation is the operation of turning a particle into its antiparticle, and vice versa. Because of the symmetry apparent in the existence of negatons and positons, positive and negative mu-mesons, protons and antiprotons, neutrons and antineutrons, it is very natural to require such an invariance property. The demand of invariance under charge conjugation puts certain restrictions on the types of terms that can appear in a general distribution function for decay.

To discuss the restrictions implied by charge conjugation we appeal to a theorem called the Lüders-Pauli theorem, or the PCT theorem.[104] If we denote the operations of space inversion, charge conjugation, and time reversal by the symbols P, C, and T, respectively, then the theorem states that for Lorentz invariant, local, Hermitian field theories (all the usual theories are included), the Lagrangian (or the field equations) is left invariant under the successive application of *all three* operations in any order, even though the theory is not invariant under any or some of the individual operations. We may summarize this by saying that the operator PCT commutes with the Hamiltonian, (PCT)H = H(PCT).

We note an important consequence of the theorem. If a theory is so formulated that it cannot be made invariant under *one* of the operations P, C, or T, it must of necessity not be invariant under at least one of the other two operations. In beta decay the fact that space inversion invariance does not hold has been established by the detection of pseudoscalar terms like $(\vec{J}\cdot\vec{p})$ and $(\vec{\sigma}\cdot\vec{p})$ in the distribution function. Since these combinations of vectors are invariant under time reversal, the PCT theorem shows that their presence implies lack of invariance under charge conjugation,[105] as well as space inversion. The presence of terms which "test" charge conjugation invariance can always be determined by observing their behavior under space inversion and time reversal, and then using the PCT theorem.

---

[103] L. D. Landau, *Nuclear Physics, 3*, 127 (1957).

[104] G. Lüders, *Annals of Physics, 2*, 1 (1957). See also Lee, Oehme, and Yang, *Physics Rev. 106*, 340 (1957).

[105] This argument depends on the fact that pseudoscalars appear with large coefficients which do not arise from final state Coulomb interactions (see preceding Section).

It is of great interest to establish experimentally whether or not time reversal invariance is violated in decay processes, along with space inversion and charge conjugation invariance. Landau,[103] Wigner,[106] and others have postulated that time reversal invariance holds, so that we can symbolically write (PC)H = H(PC) or TH = HT. The concept of inversion is then generalized to the operation PC—the mirror image of an negaton is a positon, etc. Invariance under inversion would then mean that, for example, if the *negatons* emitted in the decay of a radioactive nucleus are longitudinally *polarized anti-parallel* to their momenta, then the *positons* emitted in the decay of the corresponding *anti*-nucleus would be *polarized parallel* to their momenta. All the experiments to be discussed in the next chapter are consistent with this hypothesis of time reversal invariance, but several crucial experiments to prove it are yet to be performed (see Section 9.8).

[106] E. P. Wigner, *Rev. Mod. Phys.* *29*, 255 (1957).

CHAPTER 9

Nuclear Beta Decay

9.1. Beta Decay with Parity Conservation

The situation in nuclear beta decay prior to the discovery of the violation of space inversion symmetry will be outlined now as a preliminary to the discussion of recent work. The treatment will be brief because of the existence of the material in text books and review articles.[107] For simplicity we will concentrate on allowed transitions.

Fermi's original formulation of beta decay involved a quadrilinear combination of fermion fields:

$$H_{\text{int}} = \sum_j C_j(\bar{\psi}_p \mathcal{O}_j \psi_n)(\bar{\psi}_e \mathcal{O}_j \psi_\nu) + \text{h.c.}, \tag{9.1}$$

where ψ is a 4-component Dirac spinor and $\bar{\psi} = \psi^\dagger \beta$. With the assumption of Lorentz and space inversion invariance, the sum is over the five different covariants for the operators $\mathcal{O}_j$:

$$\mathcal{O}_j = \begin{cases} 1 & \text{Scalar (S)} \\ \gamma_\mu & \text{Vector (V)} \\ \dfrac{1}{2i}(\gamma_\mu\gamma_\nu - \gamma_\nu\gamma_\mu) & \text{Tensor (T)} \\ i\gamma_\mu\gamma_5 & \text{Axial vector (A)} \\ \gamma_5 & \text{Pseudoscalar (P)} \end{cases} \tag{9.2}$$

where the usual (Pauli) notation for the Dirac matrices is assumed. In (9.1) the appropriate scalar product between the lepton (e, ν) bracket and the nucleon (p, n) bracket is implied. For time reversal invariance the coupling constants C_j must all have the same phase (up to a sign). We lose no generality by making all the C_j real.

The usual approximation is to treat the nucleons nonrelativistically. Then the pseudoscalar interaction does not appear in "allowed" (lowest order) transitions, and there is a grouping into Fermi interactions (S and V) and Gamow-Teller interactions (T and A) with allowed nuclear matrix elements $\langle 1 \rangle$ and $\langle \vec{\sigma} \rangle$ respectively, with corresponding selection rules, $\Delta J = 0$ (no) and $\Delta J = 0, \pm 1$ (no), $0 \rightarrow 0$ forbidden.

[107] See, for instance, E. J. Konopinski and L. M. Langer, *Annual Reviews of Nuclear Science*, *2*, 261 (1953), or articles in K. Siegbahn, *Beta and Gamma Ray Spectroscopy*, North-Holland, Amsterdam, 1955.

In the standard text book manner we can find the distribution function for allowed transitions, summed over spins:

$$dw = \frac{F(Z,E)}{(2\pi)^5}\,\xi\left(1 + a\frac{\vec{p}\cdot\vec{q}}{Eq} + b\frac{m}{E}\right)pE(E_0 - E)^2\,dE\,d\Omega_e\,d\Omega_\nu\,, \quad (9.3)$$

where $(\vec{p}, E)$ is the electron momentum and energy, $(\vec{q}, q)$ is the neutrino momentum and energy, m is the electron mass, E_0 is the end-point energy, $F(Z, E)$ is the Fermi function (Coulomb correction), and ξ, a, b are constants involving the coupling constants C_i and the nuclear matrix elements $\langle 1\rangle$ and $\langle\vec{\sigma}\rangle$.

The experimental facts and their meanings in terms of the theory can be briefly summarized:

(a) Selection rules. Both $0 \rightarrow 0$ transitions, and $\Delta J = 0, \pm 1$ transitions are observed. Therefore both Fermi (S and/or V) and Gamow-Teller (T and/or A) interactions are present.

(b) Spectrum shape. The spectrum shape and ft values for the negatons and positons are consistent with the statistical factor, i.e. $b = 0$ for both $0 \rightarrow 0$ transitions and $\Delta J = \pm 1$ transitions. For Fermi[108] transitions, $b = 0 \pm 0.10$. For Gamow-Teller[109] transitions, $b = 0 \pm 0.04$. These results imply that the Fermi interaction is predominantly either S *or* V, but not both, and the Gamow-Teller interaction is either T *or* A.

(c) Absolute decay rates and relative strengths of Fermi and Gamow-Teller couplings. The beta decay of O^{14} can be used to determine the absolute magnitude of the coupling constants in the interaction (9.1).[110] The O^{14} transition is a super-allowed $0^+ \rightarrow 0^+$ transition. With the assumption of charge independence of nuclear forces the nuclear matrix element is known and the lifetime determines the coupling constant. If g_F^2 is defined as the sum of the squares of the Fermi coupling constants in (9.1), then it is found that

$$g_F = (1.41 \pm 0.01) \times 10^{-49}\ \text{erg–cm}^3\,. \quad (9.4)$$

From the systematics of all the superallowed transitions and the lifetime of the neutron (Table 4) it can be inferred that the ratio of Fermi coupling strength to Gamow-Teller coupling strength defined in a similar way is

$$\frac{g_{GT}^2}{g_F^2} = 1.3 \pm 0.3. \quad (9.5)$$

(d) (e, ν) Angular correlation. The constant a has different values for the different mixtures of interactions (for pure interactions, $a = +1$ for V, $+\frac{1}{3}$ for T, $-\frac{1}{3}$ for A, -1 for S). A study of the directional correlation between electron and neutrino (really the recoiling nucleus) for

[108] J. B Gerhart and R. Sherr, *Bull. A.P.S. 1*, 195 (1956).
[109] R. Sherr and R. H. Miller, *Phys. Rev. 93*, 1076 (1954).
[110] J. B. Gerhart, *Phys. Rev. 95*, 288 (1954); Bromley, Almquist, Gove, Litherland Paul, and Ferguson, *Phys. Rev. 105*, 957 (1957).

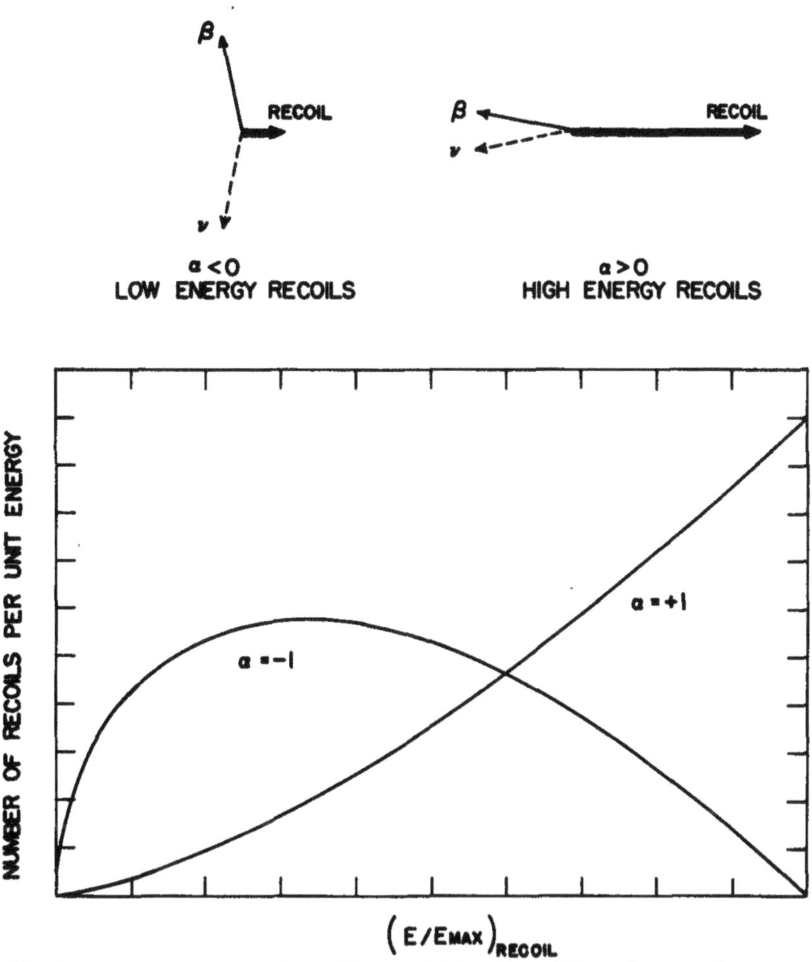

Fig. 15. Energy spectrum of recoiling nuclei for allowed beta decay. The vector diagrams indicate how the angular correlation between electron and neutrino governs the momentum of the recoiling nucleus. Low-energy recoils are expected for scalar or axial vector interactions, while high-energy recoils are characteristic of vector and tensor interactions. The curves show the energy spectrum of recoils in the extreme cases of $a = \pm 1$. For small Z and high end-point energy the recoil spectrum is $dN(x) = \frac{5}{8}x^{1/2}[3(1 - a) + (5a - 1)x]\,dx$, where $x = E/E_{\max}$ and $E_{\max} \simeq E_0^2/2M_R$.

various beta emitters will hence give information about the mixture of interactions. One technique employed is the study of the energy spectrum of the recoiling nuclei without detection of the beta particles. Fig. 15 illustrates how the shape of the energy spectrum of the recoils is connected to the electron-neutrino directional correlation. Another method is to measure the beta spectrum with a fixed angle between the beta and recoil directions. (9.3) shows that such a measurement will in principle determine a. The data are summarized in Table 9, where the nuclear beta decay is listed along with the experimental value of a, the reference to the literature, and the implications of the result. We see that the first three experiments are consistent with a beta interaction that is predominantly S and T, while the last three are consistent with predominantly V and A.

Table 9. Summary of (e, ν) Correlation Experiments

Nuclear Decay	Reference Number	Experimental Value of a	Implication
$He^6 \rightarrow Li^6 + \beta^-$ $0^+ \rightarrow 1^+$	111	$+0.34 \pm 0.12$	T (NOT A)
$n \rightarrow p + \beta^-$ $\frac{1}{2}^+ \rightarrow \frac{1}{2}^+$	112	$+0.09 \pm 0.11$	S AND T AND/OR V AND A
$Ne^{19} \rightarrow F^{19} + \beta^+$ $\frac{1}{2}^+ \rightarrow \frac{1}{2}^+$	113 114 115	-0.21 ± 0.08 $+0.14 \pm 0.2$ -0.15 ± 0.2	S AND T AND/OR V AND A
$A^{35} \rightarrow Cl^{35} + \beta^+$ $\frac{3}{2}^+ \rightarrow \frac{3}{2}^+$	116	$\begin{cases} +0.9 \pm 0.3 \\ +0.7 \pm 0.17 \end{cases}$	V (NOT S)

Before parity violations were observed, the interaction seemed to be mainly S and T. At about the time of the discovery of parity nonconservation the $A^{35}(e, \nu)$ experiment was reported. Subsequently other experiments have been performed and old ones re-evaluated. The possible existence of systematic errors in the He^6 experiment* makes that seemingly reliable evidence for a tensor interaction much less trustworthy. Better results are available† for both Ne^{19} and A^{35}, the new values being $a = 0.00 \pm 0.08$ and $a = +0.9 \pm 0.1$, respectively. Very recently the Illinois group in collaboration with workers at Argonne has examined the recoil energy spectrum from He^6 and have found‡ $a = -0.38 \pm 0.04$, in agreement with $a = -\frac{1}{3}$ for axial vector coupling, but in complete disagreement with the Rustad and Ruby value in Table 9. The results of Allen and co-workers seem especially convincing since the rather

[111] B. M. Rustad and S. L. Ruby, *Phys. Rev. 97*, 991 (1955).
[112] J. M. Robson, *Phys. Rev. 100*, 933 (1955).
[113] Maxson, Allen, and Jentschke, *Phys. Rev. 97*, 109 (1955).
[114] M. L. Good and E. J. Lauer, *Phys. Rev. 105*, 213 (1957).
[115] W. P. Alford and D. R. Hamilton, *Phys. Rev. 105*, 673 (1957).
[116] Herrmannsfeldt, Maxson, Stahelin, and Allen, *Phys. Rev. 107*, 641 (1957).

* B. M. Rustad and S. L. Ruby, post-deadline paper, New York meeting, *Am. Phys. Soc.*, Feb. 1, 1958.

† W. B. Herrmannsfeldt, Ph.D. Thesis, Physics Department, University of Illinois (1958); Herrmannsfeldt, Stahelin, and Allen, *Bull. A.P.S. 3*, 52 (1958).

‡ Herrmannsfeldt, Burman, Allen, and Braid, private communication.

different recoil spectra for He^6, Ne^{19}, and A^{35} were all obtained in the same apparatus.

These recent observations favor V and A as the covariants present in beta decay. The only complication remaining is that Ridley* has reported a value of $a = -0.05 \pm 0.10$ for Ne^{23} decay (believed to be an essentially pure Gamow-Teller transition), implying a mixture of T and A.

(e) Double beta decay. In the interaction given by (9.1) there is some arbitrariness in the treatment of the neutrinos. In the Hermitian conjugate (h.c.) term the operator for $\bar{\nu}$ (the antineutrino) replaces that for ν (the neutrino). There is a choice of whether to make the neutrino and antineutrino the same particle (Majorana neutrino) or different particles (Dirac neutrino). This arbitrariness has no consequences for ordinary beta decay, but becomes significant for processes involving more than one neutrino or inverse reactions (see item (f)). The possibility of double beta decay (a change in nuclear charge of two units without change in mass number) exists when there are two isobars differing in mass with an intervening isobar of higher mass than either of the pair (usually two even-even nuclei with an odd-odd nucleus in between). If the neutrinos of beta decay are Dirac neutrinos, this process can only occur with the emission of 4 leptons (2 electrons and 2 neutrinos), and has a very small probability beyond limits of detection even in the most favorable circumstances known. But if the neutrino and antineutrino are identical, the process can occur *without* the emission of neutrinos and has a greater (although still small) probability just within reach of observation. One can imagine the double beta decay as a two stage process with the emission of one electron and one (virtual) neutrino in the first step, and the emission of another electron and the absorption of the virtual neutrino in the second step. If ν and $\bar{\nu}$ are different this reabsorption is forbidden.

So far there is no experimental evidence for double beta decay. This was usually taken as evidence for Dirac neutrinos, although we will see later that the situation is not quite so clear cut.

(f) Inverse reactions. Another place where the distinction between Dirac and Majorana neutrinos appears is in inverse beta processes such as:

$$\begin{aligned} \bar{\nu} + p &\rightarrow n + \beta^+ \qquad \text{(a)} \\ \nu + n &\rightarrow p + \beta^- \qquad \text{(b)} \end{aligned} \tag{9.6}$$

If ν and $\bar{\nu}$ are different, the reactions (9.6) can occur, but the processes:

$$\begin{aligned} \nu + p &\rightarrow n + \beta^+ \qquad \text{(a)} \\ \bar{\nu} + n &\rightarrow p + \beta^- \qquad \text{(b)} \end{aligned} \tag{9.7}$$

are forbidden. With Majorana neutrinos the processes (9.7) can occur equally as well as processes (9.6). By ordinary first order perturbation

* B. W. Ridley, *Nuclear Physics* 6, 34 (1958).

theory we can calculate the cross section for such inverse reactions as (9.6a):

$$\sigma = \frac{1}{\pi\hbar^4 c}\frac{1}{g_\nu}\sum_{\text{spins}} |H_{\text{int}}|^2 p_e E_e, \tag{9.8}$$

where $E_e = E_\nu - 1.80$ Mev, and $g_\nu = 2$ is the statistical weight of the neutrino spins. This cross section is very small ($\sim 10^{-44}$ cm^2) and has been beyond the reach of experiment until recently. Assuming Dirac neutrinos, we note that in a nuclear reactor the beta decays are all of the type $n \rightarrow p + \beta^- + \bar{\nu}$, so that the reactor provides a source of the correct kind of neutrinos for the inverse process (9.6a), but the wrong kind for process (9.6b) occurring in the reaction ${}_{17}\text{Cl}^{37} + \nu \rightarrow {}_{18}\text{A}^{37} + \beta^-$. Reaction (9.6a) has been observed.[117] The experimental cross section, averaged over the neutrino spectrum from the reactor, is

$$\bar{\sigma}_{\text{exp}} = 4.0 \pm 1.5 \times 10^{-44}\ \text{cm}^2\,, \tag{9.9}$$

in reasonable agreement with the theoretical value[118] of 6×10^{-44} cm^2, based on $g_\nu = 2$ and the mean life of the neutron from Table 4.

R. Davis[119] has looked for the $\text{Cl}^{37} \rightarrow \text{A}^{37}$ reaction corresponding to (9.6b) with neutrinos from the Brookhaven reactor, and sets an upper limit of $0.5 \pm 0.6 \times 10^{-45}$ cm^2 to the cross section. The value expected if there were no distinction between ν and $\bar{\nu}$ is about 2.5×10^{-45} cm^2. We see that the two inverse reaction experiments support the view for Dirac neutrinos.

In summary we can say that on the basis of beta decay experiments which did not test conservation of parity, the beta interaction involved predominantly scalar and tensor covariants, or vector and axial vector, with roughly equal amounts of each type (9.5), and that the neutrino and antineutrino were different particles.

9.2. Beta Decay Interaction without Space Inversion and Time Reversal Invariance

We have seen from our general arguments in Chapter 8 that if the beta process is demanded to be invariant under space inversion and time reversal, only certain combinations of vectors can appear in the distribution functions—true scalars under space inversion and time reversal. The interaction Hamiltonian (9.1) was the most general Lorentz invariant form consistent with these requirements:

$$H_{\text{int}} = \sum_j C_j(\bar{\psi}_p \mathcal{O}_j \psi_n)(\bar{\psi}_e \mathcal{O}_j \psi_\nu) + \text{h.c.}\,, \tag{9.1}$$

where the C_j were real for time reversal invariance.

---

[117] Cowan, Reines, Harrison, Kruse, and McGuire, *Science*, *124*, 103 (1956), and *Nature*, *178*, 446 (1956).

[118] C. O. Muehlhause and S. Oleksa, *Phys. Rev.* *105*, 1332 (1957).

[119] R. Davis, *Bull. A.P.S.*, *1*, 219 (1956), and unpublished.

If the requirements of conservation of P and T (and so C, by the PCT theorem) are dropped, the interaction Hamiltonian has the following, more general form:

$$H_{\text{int}} = \sum_j (\bar{\psi}_p \mathcal{O}_j \psi_n)\{C_j(\bar{\psi}_e \mathcal{O}_j \psi_\nu) + C_j'(\bar{\psi}_e \mathcal{O}_j \gamma_5 \psi_\nu)\} + \text{h.c.}, \qquad (9.10)$$

where the constants C_j and C_j' are in general complex numbers. The presence of the $\gamma_5 = \gamma_1\gamma_2\gamma_3\gamma_4$ in the second term makes the second quadrilinear form a pseudoscalar under space inversion[120] and so violates conservation of P.

We should remark here that (9.10) with $C_j = 0$ and $C_j' \neq 0$ is completely equivalent to the old Hamiltonian (9.1), and is indistinguishable experimentally. We thus expect that there will be a maximum of parity non-conservation when $|C'| = |C|$.

The new interaction form is not the most general form that can be written. For example, the γ_5 could be put in with the nucleon fields rather than the lepton fields. The present choice is dictated by the fact that parity is a valid concept in the strong interactions, and nuclear states have definite parities. Beta decay selection rules on spin change and parity change are well established. Any alteration in the form of the nucleon bracket would destroy this agreement.

It is of some interest to see the consequences of the various operations, P, C, T on the interaction Hamiltonian. These effects can be expressed as changes in the coupling constants C_j and C_j'. The transformations are summarized in Table 10. In passing we note that the PCT theorem holds explicitly for the two types of quadrilinear Fermion interactions which appear in (9.10).

Table 10. Transformation of the Beta Decay Interaction Under the Operations P, C, and T expressed as changes in the Coupling Constants

Operation	Ordinary interaction	γ_5-type interaction	Requirement for invariance	
Unity	C	C'	—	—
Space Inversion	C	$-C'$	—	C' vanish
Charge Conjugation	C^*	$-C'^*$	C real	C' pure imaginary
Time Reversal	C^*	C'^*	C real	C' real

To illustrate the form of terms which appear when we allow the possibility of lack of invariance under P, C, T, we consider the simple case of

[120] The essential point is that γ_5 is an operator which transforms like a scalar under proper Lorentz transformations. Consequently its presence in the second term of (9.10) does not spoil the Lorentz invariance. But the space inversion operation is $P\psi(\vec{x}) = \gamma_4\,\psi(-\vec{x})$. Thus $P(\gamma_5\psi(\vec{x})) = -\gamma_5 P\psi(\vec{x})$, and there is a change of sign which makes the corresponding term in (9.10) a pseudoscalar.

allowed beta decay[121] in which we have *4* vectors or axial vectors that can appear $\vec{p}$, $\vec{q}$, $\vec{J}$, $\vec{\sigma}$. These vectors can be observed in various combinations in different experiments, some combinations being true scalars and some violating symmetry laws. We list in order of increasing experimental difficulty some of the new possibilities which violate different symmetries:

(i) $\vec{\sigma}\cdot\vec{p}$ longitudinal polarization of beta particles from non-oriented nuclei.

(ii) $\vec{J}\cdot\vec{p}$ angular distribution of beta particles from oriented nuclei.

(iii) $\vec{J}\cdot(\vec{p}\times\vec{q})$ electron and recoil angular distribution from oriented nuclei.

(iv) $\vec{\sigma}\cdot(\vec{p}\times\vec{q})$ electron and recoil angular distribution with detection of polarization of the electrons.

(v) $\vec{\sigma}\cdot(\vec{J}\times\vec{p})$ polarization and angular distribution of electrons from oriented nuclei.

The first two terms "test" P and C, but not T (except through Coulomb corrections), while the third and fourth test T and C, but not P, and the fifth term tests P and T, but not C. In the following sections we will discuss some of the experiments which have detected the first two terms, as well as others.

9.3. Directional Asymmetry in Beta Emission from Oriented Co^{60}

The now famous experiment by Wu, Ambler, Hayward, Hoppes and Hudson,[122] undertaken at the suggestion of Lee and Yang, on the directional asymmetry of the beta particles from the decay of oriented Co^{60} need hardly be described here. Independently and at the same time a group at Leyden performed a similar experiment[123] using Co^{58}. We will discuss the Co^{58} results later. Fig. 16 illustrates the directional asymmetry of the beta rays in the decay of oriented Co^{60}.

If the angular distribution is written in the form:

$$\frac{dw}{d\Omega}=\frac{1}{4\pi}\left(1+A\,\frac{\langle\vec{J}\rangle}{J}\cdot\frac{\vec{p}}{E}\right), \qquad (9.11)$$

where $\langle\vec{J}\rangle$ is the average value of the nuclear spin of the oriented Co^{60} nuclei and $(\vec{p}, E)$ is the momentum-energy of the emitted beta particle, the results found for Co^{60} implied that the energy dependence of the

---

[121] The detailed formulas will not be quoted here. The reader can refer to Jackson, Treiman, and Wyld, *Phys. Rev. 106*, 517 (1957), and also footnote 102, for the explicit coefficients of the various combinations of vectors.

[122] Wu, Ambler, Hayward, Hoppes, and Hudson, *Phys. Rev. 105*, 1413 (1957)

[123] Postma, Huiskamp, Miedema, Steenland, Tolhoek, and Gorter, *Physica 23*, 259 (1957); *24*, 157 (1958).

angular asymmetry was given by (9.11) with A independent of energy, and the magnitude of the asymmetry was the maximum possible:

$$A = -1.0 \text{ (to within 10 or 20\%)}. \tag{9.12}$$

The negaton decay of Co^{60} involves a spin change $\Delta J = -1(5^+ \rightarrow 4^+)$ and so is a pure Gamow-Teller transition. For such a transition, the coefficient A is, aside from a small Coulomb correction term:

$$A = \frac{2\,\mathrm{Re}\,(C_T C_T'^* - C_A C_A'^*)}{|C_T|^2 + |C_T'|^2 + |C_A|^2 + |C_A'|^2}. \tag{9.13}$$

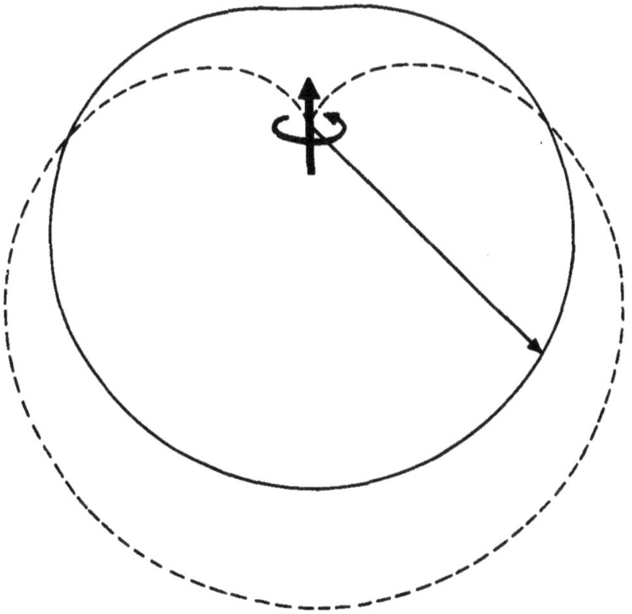

Fig. 16. Polar diagram of intensity of beta particles emitted in the decay of oriented Co^{60} nuclei. The solid vertical arrow at the center with the circulating arrow around it represents the orientation direction of the nuclear angular momentum. The beta-ray intensity is proportional to the distance from the center to the curves. The dotted curve shows the distribution of very high-energy electrons from a completely oriented sample. The solid curve is typical of the intensity for an actual experiment with moderate energy electrons and a partially oriented sample $\left(\frac{\langle J_z \rangle}{J}\frac{v_e}{c} = 0.5\right)$ Space inversion corresponds to reflecting the polar diagram in a horizontal line through the center, but leaving the direction of the angular momentum unchanged. It is apparent that this beta process is not invariant under space inversion.

Evidently A will have the value -1 for any mixture of tensor and axial vector interactions provided $C_T' = -C_T$ and $C_A' = +C_A$. Since the Gamow-Teller part of the interaction was believed to be predominantly tensor from the He^6 (e, ν) experiment (Table 9), one fixed on the choice $C_T' = -C_T$, and made the hopeful generalization that $C_j' = -C_j$ for all j.

An equally valid generalization, if the Gamow-Teller interaction should turn out to be axial vector, would be $C_j' = +C_j$. We will keep these two possibilities in mind, as well as more involved ones such as $C_T' = -C_T$, $C_A' = +C_A$.

9.4. Two-component Neutrino or its Equivalent

The fact that the nonconservation of parity occurs in the maximum possible way in the decay of Co^{60} leads one to try to find some underlying reason for it. With the choice $C' = +C$ for all the interactions, (9.10) can be rewritten in the form:

$$H_{\text{int}} = \sum_j C_j(\bar{\psi}_p \mathcal{O}_j \psi_n)(\bar{\psi}_e \mathcal{O}_j(1 + \gamma_5)\psi_\nu) + \text{h.c.}\,. \tag{9.14}$$

It is suggestive to group the $(1 + \gamma_5)$ together with ψ_ν and to attribute the violation of space inversion symmetry to some unusual property of the neutrino. Lee and Yang,[124] as well as Landau,[103] discuss this point of view as applied to μ-meson decay and to beta decay. An equally good viewpoint is that there is nothing unusual about the neutrino, but that decay interaction Hamiltonians have equal amounts of parity conserving and non-conserving parts. This latter concept allows, in principle at least, the resolution of the tau-theta puzzle where neutrinos do not appear, whereas the singling out of the neutrino as the cause of parity nonconservation does not.

Whether the factor $(1 + \gamma_5)$ appears in (9.14) because of the neutrino itself or the interaction, we can examine its effect on the states of the neutrino that are operative in beta decay and gain some insight into the mechanism of parity nonconservation. If ψ_ν is a Dirac 4-component spinor describing a neutrino of momentum $\vec{q}$ it can be written:

$$\psi_\nu = \begin{pmatrix} \varphi \\ \dfrac{\vec{\sigma}\cdot\vec{q}}{q}\varphi \end{pmatrix}, \tag{9.15}$$

where φ is a Pauli spinor describing spin up $\binom{1}{0}$ or spin down $\binom{0}{1}$. We now define a new spinor φ_ν:

$$\varphi_\nu = \tfrac{1}{2}(1 + \gamma_5)\psi_\nu. \tag{9.16}$$

Explicitly inserting ψ_ν into (9.16), we find:

$$\varphi_\nu = \frac{1}{2}\begin{pmatrix} \left(1 - \dfrac{\vec{\sigma}\cdot\vec{q}}{q}\right)\varphi \\ \left(1 - \dfrac{\vec{\sigma}\cdot\vec{q}}{q}\right)\varphi \end{pmatrix}. \tag{9.17}$$

[124] T. D. Lee and C. N. Yang, *Phys. Rev.* *105*, 1671 (1957). These authors actually use $(1 - \gamma_5)$ instead of $(1 + \gamma_5)$ since at that time the tensor interaction was favored in beta decay.

This can be called a "two-component" object since specification of the upper two components (one Pauli part) determines the other.

The spinor ψ_ν had *two* possible projections of its spin on its momentum direction. On the other hand, $\varphi_\nu \neq 0$ provided the spin is antiparallel to the momentum, but $\varphi_\nu = 0$ for spin parallel to the momentum. This means that φ_ν describes a particle with spin always antiparallel to its momentum (a left-handed screw). In a similar way, $\varphi_\nu' = \frac{1}{2}(1 - \gamma_5)\psi_\nu$ describes a particle with spin always parallel to its momentum (a right-handed screw). In the two-component theory[124] $\varphi_\nu' = \varphi_\nu^c$, that is, φ_ν describes a neutrino and φ_ν' describes an antineutrino. The important formal property is that φ_ν and φ_ν' are eigenfunctions of the operator γ_5:

$$\gamma_5\varphi_\nu = +\varphi_\nu, \qquad \gamma_5\varphi_\nu' = -\varphi_\nu'. \tag{9.18}$$

The description of the left-handed screw as neutrino and the right-handed screw as antineutrino is really arbitrary. One could equally well say that they were two different spin states of a Majorana neutrino,[125] and that in negaton emission the beta interaction was such as to cause the right screw state to appear and for positon emission the left screw state to appear.

We finally remark that if the choice $C' = -C$ were made, negaton emission would involve the emission of left-handed (anti) neutrinos, while positon emission would be accompanied by right-handed neutrinos.

9.5. Longitudinal Polarization of Beta Particles

The detection of longitudinal polarization of beta particles from non-oriented nuclei (the $\vec{\sigma} \cdot \vec{p}$ term) is a proof of violation of space inversion symmetry, just as much as the angular asymmetry found for oriented nuclei. The presence of a longitudinal polarization for beta particles emitted together with polarized neutrinos can be understood in terms of conservation of angular momentum. The simple example of $0 \rightarrow 0$ transitions is illustrated in Fig. 17. To see how this longitudinal polarization arises from the interaction itself, we consider the "two-component" form given by (9.14). The operator $(1 + \gamma_5)$ can be viewed as a projection operator which selects out a given spin state of the neutrino. It can equally well be viewed as a projection operator for the electron. If the $(1 + \gamma_5)$ is taken over to operate on the electron, the lepton bracket takes the form:

$$(\bar{\psi}_{\text{eff}}\mathcal{O}_j\psi_\nu),$$

where

$$\psi_{\text{eff}} = (1 \pm \gamma_5)\psi_e,$$

it being $(1 - \gamma_5)$ for S, T, P interactions, and $(1 + \gamma_5)$ for V, A interactions. To the extent that the electron is completely relativistic, the

[125] The equivalence of the two-component theory and the Majorana theory was pointed out by J. Serpe, *Physica, 18*, 295 (1952), and more recently by J. A. McLennan, *Phys. Rev. 106*, 821 (1957).

arguments of the preceding section apply, and we find, for example, that if the beta interaction is any linear combination of S, T, P, the high energy electrons in β^- emission will be completely polarized parallel to their line of flight.

By a simple generalization of the above argument it is easy to show that for a beta particle of velocity v, the magnitude of the longitudinal polarization is equal to v/c.

Thus, with

$$\psi_e = \begin{pmatrix} \varphi \\ \dfrac{\vec{\sigma} \cdot \vec{p}}{E + m} \varphi \end{pmatrix} \tag{9.19}$$

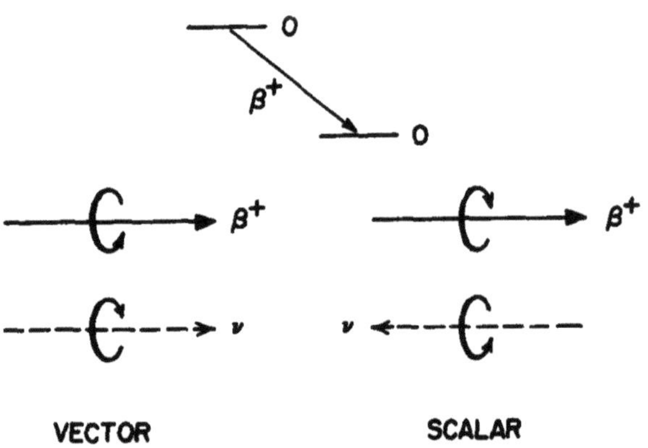

Fig. 17. Longitudinal polarization of positons in a $(0 \rightarrow 0)$ transition. Conservation of angular momentum along the line of flight of the beta particle determines the longitudinal polarization. The neutrinos accompanying the positons are assumed left-handed. For the vector interaction the positon and neutrino tend to be emitted in the same direction. Consequently the positon will be right-handed (positive longitudinal polarization). For the scalar interaction the directional correlation is opposite and so is the positon's longitudinal polarization. Experiment shows that the positons are right-handed and so favor the vector interaction with left-handed neutrinos (and/or the scalar interaction with right-handed neutrinos).

replacing (9.15), we find:

$$\psi_{\text{eff}} = \begin{pmatrix} \left(1 \mp \dfrac{\vec{\sigma} \cdot \vec{p}}{E + m}\right) \varphi \\ \mp\left(1 \mp \dfrac{\vec{\sigma} \cdot \vec{p}}{E + m}\right) \varphi \end{pmatrix}. \tag{9.20}$$

The longitudinal polarization is given by:

$$P = \frac{\left|\psi_{\text{eff}}^{\text{up}}\right|^2 - \left|\psi_{\text{eff}}^{\text{down}}\right|^2}{\left|\psi_{\text{eff}}^{\text{up}}\right|^2 + \left|\psi_{\text{eff}}^{\text{down}}\right|^2}, \tag{9.21}$$

and the computation gives $P = \mp p/E = \mp v/c$. If we have a linear combination of all five invariants and the coefficients C_j and C' are arbitrary, the longitudinal polarization is given by

$$P = Gv_e/c, \tag{9.22}$$

where, omitting Coulomb corrections:

$$G = \pm \frac{2\,\mathrm{Re}\,[|\langle 1\rangle|^2(C_S C_S'^* - C_V C_V'^*) + |\langle\vec{\sigma}\rangle|^2(C_T C_T'^* - C_A C_A'^*)]}{|\langle 1\rangle|^2(|C_S|^2 + |C_S'|^2 + |C_V|^2 + |C_V'|^2) + |\langle\vec{\sigma}\rangle|^2(|C_T|^2 + |C_T'|^2 + |C_A|^2 + |C_A'|^2)}, \tag{9.23}$$

the $\pm$ signs being appropriate for negaton and positon emission, respectively.

Frauenfelder and co-workers[126] were first to study the longitudinal polarization of the β^- particles using an electrostatic deflector to turn the longitudinal polarization into transverse polarization, which is then detected by a left-right asymmetry in ordinary Coulomb scattering (due to spin-orbit coupling forces). For Co^{60} they find the maximum polarization, $G \simeq -1$, completely consistent with the Wu-Ambler experiment on the directional asymmetry.

Subsequently other means have been used to detect longitudinal polarization. One means is Möller scattering of the beta particles by electrons aligned in a magnetized foil. Another is the observation of the circular polarization of the high-energy bremsstrahlung emitted by longitudinally polarized electrons.

Early experiments by Page and Heinberg[127] showed an appreciable polarization of the positons parallel to their line of flight in the decay of Na^{22}. This change in the sign of polarization for positons relative to negatons is just as expected from (9.23). Very rapidly longitudinal polarization data were accumulated on other Gamow-Teller transitions, on pure Fermi transitions[128] (Ga^{66} β^+, Cl^{34} β^+), on mixed transitions[129] (N^{13} β^+), as well as first-forbidden $\Delta J = 0$ (yes) transitions[130] (Au^{198} β^-), as well as many others. Within an accuracy of ~10% all data are consistent with a longitudinal polarization of the maximum amount:

$$P_{\text{exp}} \simeq \mp \frac{v}{c} \text{ for } \beta^{\mp} \text{ emission.} \tag{9.24}$$

[126] Frauenfelder, Bobone, von Goeler, Levine, Lewis, Peacock, Rossi, and DePasquali, *Phys. Rev. 106,* 386 (1957). Their result has been confirmed by other workers, for example, H. deWaard and O. J. Poppema, *Physica, 23,* 597 (1957), and Cavanagh, Turner, Coleman, Gard, and Ridley, *Phil. Mag. 2,* 1105 (1957).

[127] L. A. Page and M. Heinberg, *Phys. Rev. 106,* 1220 (1957).

[128] Deutsch, Gittelman, Bauer, Grodzins, and Sunyar, *Phys. Rev. 107,* 1733 (1957).

[129] S. S. Hanna and R. S. Preston, *Phys. Rev. 108,* 160 (1957), and Boehm, Novey, Barnes, and Stech, *Phys. Rev. 108,* 1497 (1957).

[130] Benczer-Koller, Schwarzschild, Vise, and Wu, *Phys. Rev. 109,* 85 (1958), and J. Heintze, *Z. fur Physik 150,* 134 (1958).

Examination of the expression for G in (9.23) shows that such a result demands the following choice of coupling constants:

$$C_S' = -C_S, \qquad C_T' = -C_T$$
$$C_V' = +C_V, \qquad C_A' = +C_A \tag{9.25}$$

independent of the relative amounts of the various interactions.

9.6. Further Directional Asymmetry Experiments

The directional asymmetry of beta particles from oriented nuclei has been observed for Co^{58}, Co^{56}, and the neutron, in addition to the original experiment on Co^{60}. All of these beta decays involve $\Delta J = 0$ (no), and are therefore mixed Fermi-Gamow-Teller transitions. For such transitions there is an added contribution to the coefficient A in (9.11), beyond that given by (9.13), proportional to:

$$\Delta A \sim 2\langle 1\rangle\langle\vec{\sigma}\rangle \text{ Re } (C_S C_T'^* + C_S' C_T^* - C_V C_A'^* - C_V' C_A^*). \tag{9.26}$$

The directional asymmetry in the decay of polarized neutrons has been studied at Argonne.[131] Since the matrix elements are known exactly for neutron, the value of A to be expected can be calculated if the interaction invariants are specified. With the choice of coupling constants (9.25) and the relative strengths (9.5), A is found to have the limits:

$$A = \begin{cases} -1.00 \text{ for } (S + T) \text{ and/or } (V + A) \\ -0.06 \pm 0.06 \text{ for } (S - T) \text{ and/or } (V - A) \end{cases},$$

where the coefficients were assumed real (time reversal invariance holds). For more general combinations of interactions or if time reversal invariance fails, the value of A would be between these extremes. Experimentally the coefficient is:

$$A_{\text{exp}} = -0.09 \pm 0.03. \tag{9.27}$$

This value is consistent with (S − T) and/or (V − A), and shows that the Fermi-Gamow-Teller cross term (9.26) is present in neutron decay. Observation of a large positive value for B (9.31), the coefficient of the correlation between *neutrino* direction and neutron spin, implies (V − A) rather than (S − T).[131]

The directional asymmetry of the positons from oriented Co^{58} has been observed at both Leyden[123] and the National Bureau of Standards.[132] The result for the asymmetry parameter is:

$$A = +0.33 \pm 0.05, \tag{9.28}$$

just the value one would expect[133] for a pure Gamow-Teller *positon* emitter

[131] Burgy, Krohn, Novey, Ringo, and Telegdi, *Phys. Rev. 110*, 1214 (1958).
[132] Ambler, Hayward, Hoppes, and Hudson, *Phys. Rev. 108*, 503 (1957).
[133] The difference of a factor of 3 from the asymmetry for Co^{60} is due to a geometrical factor depending on the spins and spin change.

with a transition $2^+ \rightarrow 2^+$. The absence of the added contribution to A given by (9.26) would indicate that the Fermi matrix element was small (perhaps less than one tenth) compared to the Gamow-Teller matrix element. This explanation is contradicted by an experiment[134] on the angular distribution of the gamma rays subsequent to the decay of aligned Co^{58}, which experiment implies that $|\langle 1 \rangle| \simeq \frac{1}{3}|\langle \vec{\sigma} \rangle|$, yielding $A \simeq +0.8$ or -0.2 with assumptions similar to those made above for the neutron decay. Other reasons for the smallness of the interference term (9.26) can be advanced, such as lack of time reversal invariance, or an interaction that is predominantly (S, A) or (V, T), but such explanations contradict the neutron experiment as well as the (e, ν) correlation experiments of Section 9.1.

A similar directional asymmetry experiment has been done with Co^{56}, yielding an asymmetry parameter:[132]

$$A = +0.22 \pm 0.02, \tag{9.29}$$

consistent with a pure Gamow-Teller positon emitter with spin change $4^+ \rightarrow 4^+$. In this case there is no independent evidence on the actual magnitude of the Fermi and Gamow-Teller matrix elements.

9.7. Polarization Effects following Beta Decay

We have seen in previous sections how the violation of space inversion symmetry gives rise to certain polarization effects and directional correlations. In particular, the longitudinal polarization of beta particles and the directional asymmetry of beta emission from oriented nuclei were discussed. In the decay of non-oriented nuclei other polarization effects arise. For example, the directional asymmetry expression (9.11) can be reinterpreted to mean that the recoiling nucleus is partially polarized along the direction of emission of the beta particle. The detection of this polarization will give independent information[135] on the coefficient A. One method of detection is observation of the circular polarization of subsequent gamma radiation in directional correlation with the beta particle.[136] Several experiments of this type have been performed,[137] with results consistent with the assignment of coupling constants given by (9.25). In particular, an experiment[138] on Sc^{46} (a mixed $\Delta J = 0$ transition) indicates the presence of the Fermi-Gamow-Teller interference term ΔA (Eq. (9.26)) with appreciable magnitude.

---

[134] D. F. Griffing and J. C. Wheatley, *Phys. Rev. 104*, 389 (1956).

[135] Actually, a slight reinterpretation of the symbols in A is necessary. See the discussion in the reference of footnote 139.

[136] Alder, Stech, and Winther, *Phys. Rev., 107*, 723 (1957).

[137] H. Schopper, *Phil. Mag. 2*, 710 (1957); H. Appel and H. Schopper, *Z. fur Physik, 149*, 103 (1957); F. Boehm and A. H. Wapstra, *Phys. Rev. 107*, 1462 (1957).

[138] F. Boehm and A. H. Wapstra, *Phys. Rev. 107*, 1202 (1957); Lundby, Patro, and Stroot, *Nuovo Cimento, VII*, 891 (1958).

The beta gamma circular polarization correlation involves the same coefficient A as the directional asymmetry of beta particles from oriented nuclei, but other polarizations involve new combinations of coupling constants. In K capture or beta emission the residual recoiling nucleus is polarized along its line of flight.[139] For K capture the polarization is proportional to a coefficient B in the notation of footnote 121, while in beta emission it is proportional to $(A + B)$. The important point is that, assuming the relations (9.25) are valid, the directional asymmetry coefficient A is proportional to

$$A \sim |C_A|^2 + |C_T|^2, \tag{9.30}$$

while the coefficient B is proportional to

$$B \sim |C_A|^2 - |C_T|^2. \tag{9.31}$$

Consequently, whereas A involves the sum of tensor and axial vector contributions and cannot be used to distinguish between them, B involves the difference and can be used to discriminate between the two couplings. Detection of the longitudinal polarization of the recoils relative to their line of flight, therefore, yields new information on the type of interaction akin to the information from electron-neutrino correlation experiments. These recoil polarizations occur only in pure Gamow-Teller and mixed transitions. In K capture, however, there is a polarization of the hole left in the K shell relative to the recoil direction even for pure Fermi transitions. This hole polarization is proportional to

$$H \sim |\langle 1 \rangle|^2(|C_S|^2 - |C_V|^2) + \frac{|\langle \vec{\sigma} \rangle|^2}{3}(|C_A|^2 - |C_T|^2), \tag{9.32}$$

and so gives rise to the possibility of discriminating between the scalar and vector couplings, as well as tensor and axial vector.

The detection of the recoil polarization can be accomplished by observation of the circular polarization of subsequent gamma radiation correlated with the recoil direction, or the corresponding longitudinal polarization of conversion electrons.[139] The polarization of the hole in the K shell after K capture can perhaps be detected by a similar study of the circular polarization correlation of the X-rays emitted after K capture.

Recently an ingenious experiment measuring the recoil-gamma circular polarization correlation following K-capture in Eu^{152m} has been reported.[140] The relevant parts of the decay scheme and a pictorial representation of the effect are shown in Fig. 18. The preferential detection of gamma rays emitted along the recoil direction is accomplished by use of a resonance fluorescence detector which is sensitive only to those photons whose Doppler shift due to motion of the excited recoiling nucleus almost compensates for the sharing of the transition energy between the gamma

[139] Frauenfelder, Jackson, and Wyld, *Phys. Rev. 110*, 451 (1958).
[140] Goldhaber, Grodzins, and Sunyar, *Phys. Rev. 109*, 1015 (1958).

ray and the final nucleus. The results indicate that the neutrinos emitted are left-handed. Coupled with the evidence from Sections 9.3 and 9.5, this experiment shows that the dominant Gamow-Teller interaction is *axial vector*, in agreement with the new (e, ν) correlation experiment on He^6 (Section 9.1) and the neutron experiment (Section 9.6).

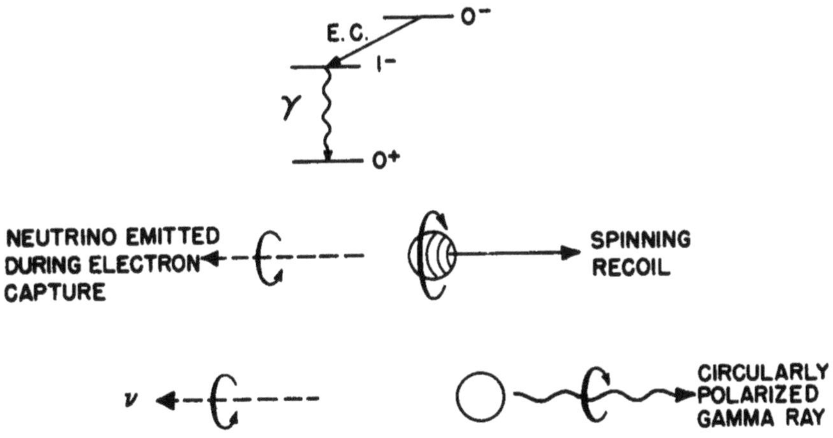

Fig. 18. Correlation between neutrino spin state in electron capture and circular polarization of a subsequent gamma ray emitted along the line of flight of the recoil. The decay scheme is a simplified version of that for Eu^{152m} (see Footnote 140). Conservation of angular momentum along the line of flight demands that the gamma ray have a circular polarization of the same sense as the longitudinal polarization of the neutrino. For Eu^{152m} decay the circular polarization is negative, corresponding to left-handed neutrinos and the axial vector interaction.

9.8. Time Reversal Experiments

As discussed in Section 8.3, experimental tests of time reversal invariance can be made by looking for the presence of triple scalar products of momentum or angular momentum vectors in beta decay distribution functions, or perhaps by studying Coulomb corrections to the ordinary scalar product terms. The electron-neutrino correlation from oriented nuclei[121] $\vec{J} \cdot (\vec{p} \times \vec{q})$, the beta-gamma directional correlation from oriented nuclei[141] $(\vec{J} \cdot (\vec{p} \times \vec{k})(\vec{J} \cdot \vec{k}))$, and the beta transverse polarization-gamma correlation from unoriented nuclei[142] $(\vec{\sigma} \cdot (\vec{p} \times \vec{k})(\vec{p} \cdot \vec{k}))$ are three possibilities.

A beta-gamma directional correlation from oriented Co^{58} was looked for,[132] but none was found. The coefficient of the $\vec{J} \cdot (\vec{p} \times \vec{k})(\vec{J} \cdot \vec{k})$ term is of the same form as ΔA (see (9.26)), except that it involves the imaginary part of the same products of coupling constants rather than the real part. In view of the uncertainties in the interpretation of the Co^{58} beta

[141] R. B. Curtis and R. R. Lewis, *Phys. Rev. 107*, 1381 (1957); M. Morita and R. S. Morita, *Phys. Rev. 107*, 1316 (1957).

[142] R. B. Curtis and R. R. Lewis, *Phys. Rev. 107*, 543 (1957).

directional asymmetry experiment, nothing conclusive can be said about time reversal invariance from this experiment. A series of experiments* on oriented Mn^{52} nuclei, studying the beta directional asymmetry as well as beta-gamma correlations, have shown that the phase angle between the Fermi and Gamow-Teller coupling constants lies between 140° and 250°. This is consistent with (V − A) and/or (S − T), as found from the neutron experiment described in Section 9.6, but still allows appreciable lack of time reversal invariance.

Experiments on the electron-neutrino correlation from oriented neutrons are in progress at Chalk River and Argonne. Preliminary results from Chalk River† give a value $D = -0.02 \pm 0.28$ for the coefficient of $\vec{J} \cdot (\vec{p} \times \vec{q})/Eq$ in the neutron decay distribution,[121] compared to a maximum absolute value of ~ 0.5 with the choice of constants (9.5) and (9.25) and a phase of 90° between the Fermi and Gamow-Teller parts. The neutron and Mn^{52} experiments, while lacking high precision, are consistent with time reversal invariance, and at least show that in beta decay there is no spectacular violation of this symmetry principle. It seems likely that the law of PC invariance proposed by Landau, Wigner, and others holds in beta decay and other decay processes (see Section 8.4).

9.9. Reinterpretation of the "Old" Experiments

In Section 9.1 we presented a survey and interpretation of the experiments done largely before parity nonconservation was discovered. These experiments must now be re-evaluated in the light of our knowledge about the violation of space inversion symmetry. The selection rules, decay rates, and electron-neutrino correlations are unaffected by the generalization implied by (9.10). But the vanishing of the Fierz terms (the coefficient b in (9.3)), which previously indicated S *or* V, and T *or* A, now proves essentially nothing. The only conclusion is that

$$\mathrm{Re}\,(C_S C_V^* + C_S' C_V'^*) \simeq 0 \simeq \mathrm{Re}\,(C_T C_A^* + C_T' C_A'^*).$$

The choice (9.25), for example, makes these terms vanish regardless of the relative amounts of S and V, or T and A.

The problem of double beta decay (Section 9.1 (e)) is complicated. If the interaction (9.10) has $C' = +C$, as in (9.14), or $C' = -C$, the new situation is equivalent to the Dirac neutrino case of before, in agreement with experiment. But if the coefficients C and C' are not so simply related, there is the possibility of double beta decay at a rate intermediate between the "yes or no" of the Majorana and Dirac neutrino theories—something that requires quantitative experimental investigation.[143]

The interpretation of the inverse reactions (Section 9.1(*f*)) is modified in

* Ambler, Hayward, Hoppes, and Hudson, *Phys. Rev. 110*, 787 (1958).

† Clark, Robson, and Nathans, paper submitted to the Second International Conference on the Peaceful Uses of Atomic Energy, Geneva, September, 1958.

[143] This has bearing on the problem of conservation of leptons. See Section 10.4.

an interesting manner. If the "two-component" theory holds, the (anti)-neutrinos emerging from a reactor all have their spins aligned along their momenta[144] in just the correct posture for absorption, according to (9.6a). This means that in (9.8) the neutrino statistical weight factor is $g_\nu = 1$, instead of the old value of 2. Hence the theoretically expected cross section is about 12×10^{-44} cm^2 for the Cowan and Reines experiment, and 5×10^{-45} cm^2 for the Davis experiment. The observed value of 4×10^{-44} cm^2 in the Cowan-Reines experiment is now a factor of 3 smaller than the calculated value.

9.10. A Possible Beta Interaction

From the previous sections we can draw certain important conclusions about the beta interaction, while at the same time we perhaps face contradictions within the framework of the usual theory. The beta longitudinal polarization experiments, as well as the directional asymmetries from oriented nuclei, indicate that the odd and even parts of the interaction Hamiltonian have equal magnitudes. If V and/or A are present, these experiments demand $C'_{\mathrm{V,A}} \simeq +C_{\mathrm{V,A}}$. If the interaction involves S and/or T, then $C'_{\mathrm{S,T}} \simeq -C_{\mathrm{S,T}}$. The electron-neutrino correlation experiments (Table 9) yield conflicting information, but seem to exclude such combinations as (V, T) or (S, A). The simplest conclusion is that the interaction is predominantly either S and T or V and A, although other combinations are possible.

The recent experiment by Goldhaber and coworkers[140] on K capture indicates A as the dominant Gamow-Teller invariant. Although this result depends to some extent on the assignment (9.25) of the coupling constants from electron polarization experiments, and is in that sense a derived conclusion, we will tentatively assume that the nuclear beta interaction is predominantly V and A with $C' = C$. The interaction Hamiltonian will be of the form:

$$H_{\mathrm{int}} = \sum_{\mathrm{V,A}} C_j(\bar{\psi}_p \mathcal{O}_j \psi_n)(\bar{\psi}_e \mathcal{O}_j (1 + \gamma_5)\psi_\nu) + \text{h.c.}\,. \tag{9.33}$$

The directional asymmetry for the neutron (Section 9.6) then implies that the linear combination is (V − A). The neutron lifetime and super-allowed ft values (Section 9.1(c)) indicate $|C_{\mathrm{A}}/C_{\mathrm{V}}| \simeq 1.14 \pm 0.14$, with $|C_{\mathrm{V}}|^2$ given by $(g_F^2/2)$ of (9.4). This interaction is consistent with all the "parity" experiments, except the directional asymmetry measurements on Co^{58} (see Section 9.6), and is in agreement with all the older observations except the original electron-neutrino correlation experiment on He^6 (Table 9), now discredited. In view of the difficulty of some of the measurements involved, it is perhaps justifiable to stress the accord between (9.33) and experiment, and to ignore the disagreements.

[144] Either parallel, if we assume $C' = C$, or antiparallel if we assume $C' = -C$. Which way is not relevant since the emission and absorption is assumed to be governed by the same interaction.

CHAPTER 10

Other Decay Processes

The Fermi mechanism for beta decay seems to apply also to the decay of the μ-meson, as well as to μ-meson capture by nuclei. The appearance of even and odd parts in the Hamiltonian (9.10) for nuclear beta processes, with all the polarization and correlation effects going along with them, would imply that similar phenomena should appear in μ-meson decay and capture. Indeed, Lee and Yang suggested in their original paper that study of μ-meson decay might yield information on nonconservation of parity. In the present chapter we first discuss μ decay and then briefly consider μ capture. The similarity in the absolute strengths of the two μ-meson Fermi interactions and that of nuclear beta decay leads to a discussion of the question of a universal Fermi interaction for all decay processes, as well as the concept of conservation of leptons. The decay mechanism for bosons (pions and K-mesons) is then described in a qualitative way, as is the mechanism of hyperon decay. The discussion of all these topics, especially of the K-meson and hyperon decays, is rather brief since there is an excellent review article available.[145]

10.1. Mu Meson Decay

The μ-meson decays into an electron and two neutrinos, as indicated in Table 1, with an electron energy spectrum typical of a relativistic three-body decay. The maximum electron energy is about 52 Mev. Some energy distributions expected from theory are shown in Fig. 19.

Following the suggestions of Lee and Yang and the observation of parity nonconservation in beta decay by Wu et al, Garwin, Lederman, and Weinrich,[146] in a dramatic and ingenious experiment, observed a directional asymmetry in the emission of the electrons in μ decay relative to the line of flight of the μ-meson. Independently, Friedman and Telegdi[147] observed a similar asymmetry from the decay of μ-mesons arising from pions stopped in emulsions. The effects observed in μ decay are of the same sort as those first observed for oriented Co^{60} (Section 9.3). They indicate violation of space inversion symmetry not only in μ decay but in pion

[145] M. Gell-Mann and A. H. Rosenfeld, *Annual Reviews of Nuclear Science*, *7*, 407, (1957).

[146] Garwin, Lederman, and Weinrich, *Phys. Rev. 105*, 1415 (1957).

[147] J. I. Friedman and V. Telegdi, *Phys. Rev. 105*, 1681 (1957), and *Phys. Rev. 106*, 1290 (1957).

decay as well, since the μ-meson must have been at least partially polarized in the pion decay process.

The decay process can be described in terms of an interaction Hamiltonian similar to the nuclear beta decay Hamiltonian. Since the nuclear experiments imply that the odd and even parts of the interaction are

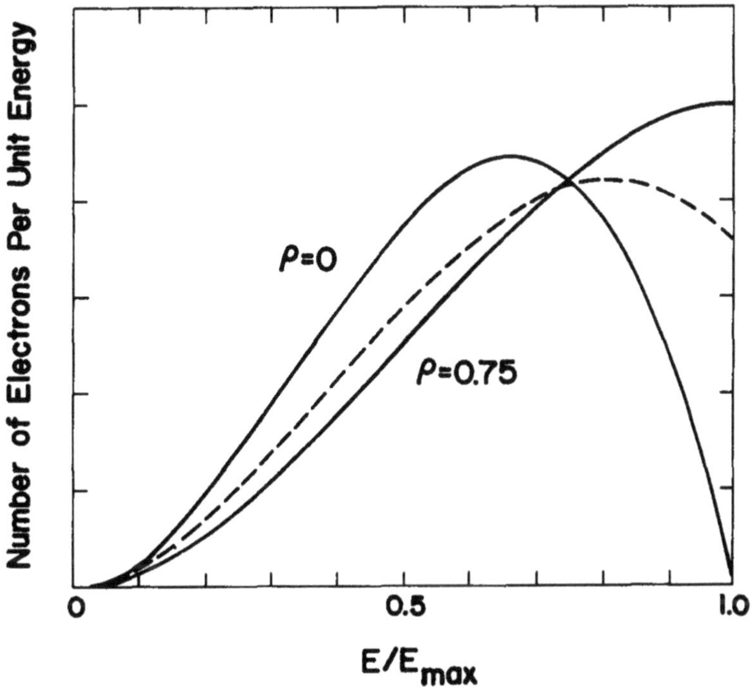

Fig. 19. Energy spectrum of electrons emitted in the decay of the μ-meson. The maximum energy is 52 Mev. Curves for the Michel parameter $\rho = 0$ and $\rho = 0.75$ are shown, as well as a dotted curve representing the statistical phase space distribution. Experiments indicate a spectrum with $\rho = 0.68 \pm 0.02$.

present in equal amounts, it is natural to discuss the μ decay problem in terms of "two-component" neutrinos or their equivalent. Consequently the interaction will be written in the form:

$$H_{\text{int}} = \sum_{\text{V,A}} f_j(\bar{\psi}_e \mathcal{O}_j \psi_\mu)(\bar{\varphi}_\nu \mathcal{O}_j \varphi_\nu) + \text{h.c.}\,, \tag{10.1}$$

where the two neutrinos have been grouped together for calculational convenience.[148] The neutrinos represented by φ_ν are not ordinary 4-component Dirac spinors. In fact, $\varphi_\nu = \frac{1}{2}(1 + \gamma_5)\psi_\nu$, where ψ_ν is a conventional 4-component spinor. φ_ν is an eigenfunction of γ_5:

$$\gamma_5 \varphi_\nu = \varphi_\nu, \tag{10.2}$$

[148] The ordering of the various wave functions is not of fundamental significance since a re-ordering merely gives a different linear combination of invariants.

and corresponds to a left-handed neutrino, as is appropriate for the beta interaction of Section 9.10. The sum in (10.1) includes only two terms for V and A, since the presence of the two φ_ν factors makes the contributions of the other invariants vanish identically.[149] We note, in passing, the similarity between the μ decay interaction (10.1) and the beta interaction (9.33), a point to which we will return below.

The interaction form in (10.1) describes the decay process:

$$\mu^\pm \to e^\pm + \nu + \bar{\nu}, \tag{10.3}$$

but not

$$\mu^\pm \to e^\pm + \nu + \nu$$

or (10.4)

$$\mu^\pm \to e^\pm + \bar{\nu} + \bar{\nu}.$$

One observable difference between the emission of one right- and one left-handed neutrino and the emission of two right- (or left-) handed ones is the shape of the energy spectrum of the electrons. Michel[150] showed that the energy spectrum of the decay electrons had the general form:

$$N(x)\,dx = 4x^2[3(1 - x) + \tfrac{2}{3}\rho(4x - 3)]\,dx,$$

depending on the single parameter ρ. The variable x is the electron energy in units of its maximum value ($x = 2E/m_\mu$). The decay process (10.3) with the emission of neutrinos with different spin states has a theoretical ρ value, $\rho_{\text{theo}} = 0.75$, while the processes (10.4) both have $\rho_{\text{theo}} = 0$. The experimental value[151] of the Michel parameter is $\rho_{\text{exp}} = 0.68 \pm 0.02$. While the agreement between $\rho = 0.75$ and 0.68 is not perfect, the decay processes (10.4) seem definitely ruled out. We will assume that the decay is according to (10.3) and described by the Hamiltonian (10.1).

It is a straightforward matter to calculate the energy and angular distribution of the decay electron from a polarized μ-meson:

$$dw = \frac{1}{2\pi}\, x^2\, [(3 - 2x) \mp \xi(1 - 2x)\cos\theta]\, dx\, d\Omega, \tag{10.5}$$

where

$$\xi = \frac{f_V f_A^* + f_A f_V^*}{|f_V|^2 + |f_A|^2}\,. \tag{10.6}$$

The signs in (10.5) are appropriate to negative and positive μ-meson decay, respectively. The spherically symmetric term in (10.5) gives the ordinary energy spectrum with $\rho = 0.75$. The angle θ is measured relative to the

---

[149] With the substitution $\varphi_\nu = \frac{1}{2}(1 + \gamma_5)\psi_\nu$, the neutrino bracket becomes $\frac{1}{4}(\bar{\psi}_\nu(1 - \gamma_5)\mathcal{O}_j(1 + \gamma_5)\psi_\nu)$. If $\mathcal{O}_j$ commutes with γ_5 (S, T, P), this vanishes identically, while if $\mathcal{O}_j$ anticommutes with γ_5 (V, A), it becomes $\frac{1}{2}(\bar{\psi}_\nu\mathcal{O}_j(1 + \gamma_5)\psi_\nu)$.

[150] L. Michel, *Proc. Phys. Soc. A 63*, 514, 1371 (1950).

[151] Sargent, Rinehart, Lederman, and Rogers, *Phys. Rev. 99*, 885 (1955); L. Rosenson, *Phys. Rev. 109*, 958 (1958); K. M. Crowe (unpublished).

spin direction of the decaying meson. The angular distribution, averaged over the electron energy distribution, is:

$$\left\langle\frac{dw}{d\Omega}\right\rangle_E = \frac{1}{4\pi}\left(1 \pm \frac{\xi}{3}\cos\theta\right). \tag{10.7}$$

After the original work many different groups working with both emulsions and electronic techniques have studied the angular asymmetry in μ decay, mainly from positive μ-mesons. The various experiments[152] imply a value for ξ:

$$|\xi| = 0.87 \pm 0.13. \tag{10.8}$$

The sign is not known from these experiments because, although the electrons are emitted preferentially backwards relative to the line of flight of the μ-meson (for both positive and negative mesons), the spin direction of the μ-meson is not known. The observed energy dependence of the directional asymmetry[153] is reasonably consistent with (10.5).

The sign of ξ can be determined by observing the longitudinal polarization of the decay electrons. By an argument essentially identical with that given in Section 9.5 for the longitudinal polarization of beta particles, we find that for randomly oriented μ-mesons, the longitudinal polarization of the decay electrons is:

$$P = \pm\xi\frac{v}{c} \simeq \pm\xi. \tag{10.9}$$

Again the $\pm$ signs refer to negative and positive μ-meson decay respectively. Experiments at Liverpool[154] have showed that the positons from μ^+ decay are polarized *parallel* to their momentum, indicating that ξ is negative.

Combining the longitudinal polarization result with the asymmetry data, we find that $f_A \simeq -f_V$, or that the interaction is (V — A) in the ordering of spinors that appears in (10.1). The marked similarity between this decay interaction and the beta decay interaction (9.33) is obvious. There remains, of course, the ambiguity about the ordering of the four spinors and the question of which ordering in μ decay should be compared to the beta decay ordering. Since the beta interaction involves a lepton (light fermion) bracket $(\bar{e}\nu)$, it is perhaps reasonable to rewrite the μ decay quadrilinear form as $(\bar{e}\mu)(\bar{\nu}\nu) \rightarrow (\bar{\nu}\mu)(\bar{e}\nu)$ for comparison of the two interactions. It turns out that the combination (V — A) is unchanged by this reordering process. Consequently we find that the interactions

[152] D. H. Wilkinson, *Nuovo Cimento, VI*, 516 (1957) presents a survey of the various experiments and determines the averaged value of $|\xi|$.

[153] Berley, Coffin, Garwin, Lederman, and Weinrich, *Phys. Rev. 106*, 835 (1957).

[154] Culligan, Frank, Holt, Kluyver, and Massam, *Nature, 180*, 751 (1957).

are essentially identical in form.[155] In Section 10.4 we will see that the neutron and μ-meson decay rates imply that the beta decay and μ-meson decay coupling constants are very similar in magnitude, the difference of a factor of 10^9 in lifetime being due to the great difference in energy release.

Our discussion of μ decay has been based on the simple local quadrilinear interaction of (10.1). There are a number of points which indicate that this description may be oversimplified. There is first of all the question of the ρ value. Experimentally, $\rho = 0.68 \pm 0.02$, while the theory gives $\rho = 0.75$. Secondly, ξ is not exactly equal to unity, and the departure from unity may be significant. Modifications of the theory can be made in several directions:

(a) To change the ρ value, mix in some of the processes (10.4) along with (10.3).

(b) To reduce ξ and change ρ, abandon the simple "two-component" model and treat the decay with 20 arbitrary constants.[156] This introduces two parameters into the theory in addition to ξ (ρ and one other).

(c) Abandon the specifically local character of the Fermi interaction.[157] The last possibility is perhaps the most attractive one. It may well be that nonlocal effects could occur in μ decay where the energy release is large, but not appear in nuclear beta decay because of the much lower energy differences involved.

10.2. Mu Meson Capture

The reaction in which a bound negative μ-meson is captured by a proton in a nucleus with the emission of a neutrino:

$$\mu^- + p \rightarrow n + \nu \tag{10.10}$$

is called μ-meson capture and is the μ-mesonic equivalent of orbital electron capture in nuclear beta decay. The energy release is about 104 Mev. The interaction can be described in terms of a Hamiltonian of the

[155] It should be remarked that the nuclear beta interaction (9.33) is based on conflicting experimental data. It is possible that the interaction is mainly (S-T) with $(1 - \gamma_5)$ multiplying the neutrino spinor. Then in μ decay it would be natural to use right-handed neutrinos, rather than left-handed. The results are the same as given here, except that in (10.5), (10.7), and (10.9) the signs of ξ are opposite. The experiments then imply (V + A), rather than (V − A). Under the reordering $(\bar{e}\mu)(\bar{\nu}\nu) \rightarrow (\bar{\nu}\mu)(\bar{e}\nu)$ the linear combination (V + A) goes into a new linear combination (S-P). This can be thought essentially the same as the nuclear beta interaction since the presence or absence of P is not known, and T can be added formally, although giving no contribution in μ decay. The possibility discussed in this footnote is less desirable from a theoretical viewpoint since it implies violation of conservation of leptons in pion decay (See Section 10.4).

[156] C. Bouchiat and L. Michel, *Phys. Rev. 106*, 170 (1957), and Larsen, Lubkin, and Tausner, *Phys. Rev. 107*, 856 (1957).

[157] T. D. Lee and C. N. Yang, *Phys. Rev. 108*, 1611 (1957), and S. Bludman and A. Klein, *Phys. Rev. 109*, 550 (1958).

form of (9.1) or (9.10) with the electron replaced by a μ-meson. Negative μ-mesons stopping in matter very rapidly slow down and get captured into *1s* atomic orbits around nuclei. The nuclear capture process can accordingly be treated in complete analogy with the K capture of orbital electrons. It is easy to show that the capture rate is given by the approximate expression:

$$\frac{1}{\tau_c} \simeq \frac{g_{\mu N}^2}{\pi} Z|M|^2 E_\nu^2 \left(\frac{Z^3}{\pi a_\mu^3}\right) . \tag{10.11}$$

Here $g_{\mu N}^2$ is the effective coupling constant for the μ-meson-nucleon Fermi interaction,[158] E_ν is the neutrino energy and a_μ is a μ-mesonic Bohr radius (2.55×10^{-11} cm). The factor $(Z^3/\pi a_\mu^3)$ is the density of μ-meson's K-shell wave function at the nucleus. Due to the finite nuclear size, this factor is a good approximation only for small Z.[159] The product $Z|M|^2$ represents the effective number of protons available for the capture process. If the meson is captured by a proton at rest, the resulting neutron recoils with a kinetic energy of about 5 or 6 Mev. Inside a nucleus the Pauli principle will forbid such a recoil unless the neutron goes into a state outside the already occupied states. In terms of a Fermi gas model, the recoil momentum is about 100 Mev/c compared to the radius of the Fermi sphere of roughly 200 Mev/c. A simple geometrical calculation indicates that the exclusion principle inhibition makes $|M|^2 \sim \frac{1}{4}$. Some excitation energy ($\sim$10 or 20 Mev) is imparted to the nucleus, but the major part of the μ-meson's rest energy goes to the neutrino. As a crude estimate we may take $E_\nu \simeq m_\mu$. With these estimates we obtain the rough approximation:

$$\frac{1}{\tau_c} \simeq \frac{1}{4\pi^2} g_{\mu N}^2 e^6 m_\mu^5 Z^4 . \tag{10.12}$$

The capture rate given by (10.12) is a mode of decay added to the natural decay rate of the μ-meson. Consequently, the observed decay rate for bound negative μ-mesons is:

$$\frac{1}{\tau} = \frac{1}{\tau_\mu} + \frac{1}{\tau_c} , \tag{10.13}$$

where τ_μ is the mean lifetime of the *free* μ-meson (2.22×10^{-6} sec). The rapid Z-dependence of the capture rate (10.12) means that the observed lifetime decreases rapidly with increasing atomic number of the stopping material. It is convenient to define the atomic number Z_0 as the atomic

[158] In rough approximation, $g_{\mu N}^2$ is proportional to $(g_F^2 + 3g_{GT}^2)_{\mu N}$. See H. Primakoff, *Proceedings of the Fifth Annual Rochester Conference on High Energy Physics* Interscience, New York, 1955.

[159] J. A. Wheeler, *Rev. Mod. Phys. 21*, 133 (1949).

number for which the observed lifetime is one half of the free lifetime ($\tau_c = \tau_\mu$). Then the strength of coupling can be written:

$$g_{\mu N}^2 \simeq \frac{4\pi^2}{Z_0^4 \tau_\mu e^6 m_\mu^5}\left(\frac{\hbar^{10}}{c}\right). \tag{10.14}$$

In (10.14) we have restored the proper powers of $\hbar$ and c to make the result dimensionally correct. Experimentally it is found that $Z_0 \simeq 11$. This leads to the value:

$$g_{\mu N} \simeq 1.5 \times 10^{-49} \text{ erg-cm}^3 . \tag{10.15}$$

10.3. Universal Fermi Interaction

The striking similarity in the absolute magnitudes of the μ-capture coupling constant $g_{\mu N}$ (10.15) and the Fermi beta decay constant g_F (9.4) leads to the enquiry as to whether there is some common coupling between the various groups of fermions. We first review the derivation of the magnitudes of the coupling constants for nuclear beta decay and μ-meson decay.

For allowed transitions the relationship between the decay rate and the coupling constants is:

$$g_F^2|\langle 1\rangle|^2 + g_{GT}^2|\langle\vec{\sigma}\rangle|^2 = \frac{2\pi^3(0.693)}{m^5 \text{ ft}}\left(\frac{\hbar^7}{c^4}\right). \tag{10.16}$$

where m is the electron mass and ft is the usual product of the integral over the spectrum times the half-life. For O^{14} the Fermi matrix element is $|\langle 1\rangle|^2 = 2$, while ft $= 3088 \pm 56$ sec.[110] This yields the result (9.4) for the Fermi coupling constant g_F. For the neutron, $|\langle 1\rangle|^2 = 1$, $|\langle\vec{\sigma}\rangle|^2 = 3$, and the lifetime in Table 4 implies ft $= (1.2 \pm 0.2) \times 10^3$ sec. If we define an average beta decay coupling strength g_β:

$$g_\beta^2 = \frac{1}{4}(g_F^2 + 3g_{GT}^2) ,$$

then the neutron data indicate

$$g_\beta = (1.58 \pm 0.16) \times 10^{-49} \text{ erg-cm}^3 . \tag{10.17}$$

Since the O^{14} data are more accurate we will take g_F (9.4) as the beta interaction strength, disregarding the possible difference between g_F and g_{GT}.

In the decay of the μ-meson the coupling strength and decay rate are related by:

$$g_{\mu e}^2 = \frac{384\ \pi^3}{m_\mu^5 \tau_\mu}\left(\frac{\hbar^7}{c^4}\right), \tag{10.18}$$

where

$$g_{\mu e}^2 = \frac{1}{2}(|f_V|^2 + |f_A|^2) \tag{10.19}$$

is the average coupling constant squared in the notation of Section 10.1. With the observed lifetime of $2.22 \pm 0.02 \times 10^{-6}$ sec, the coupling strength is

$$g_{\mu e} = (2.01 \pm 0.01) \times 10^{-49} \text{ erg-cm}^3 . \tag{10.20}$$

The close agreement between the absolute magnitudes of the coupling constants for beta decay (9.4) or (10.17), μ-meson decay (10.20), and μ capture (1.015) led to the idea that the nucleons and leptons (μ-mesons, electrons, and neutrinos) could be coupled together in a universal Fermi interaction[160] symbolized by the triangle in Fig. 20. It was postulated

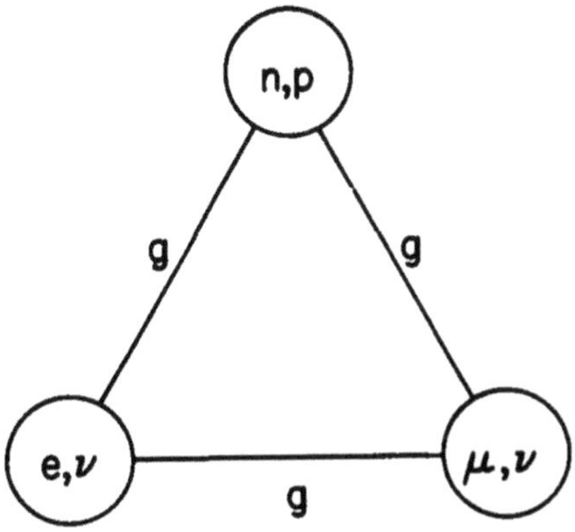

Fig. 20. The Tiomno-Wheeler triangle for a universal Fermi interaction between the nucleons and the light fermions. The coupling strength in each interaction is g.

that the coupling strengths and the detailed forms of the interaction were the same for each arm of the triangle. The essential identity in form between the μ decay Hamiltonian (10.1) and the beta decay Hamiltonian (9.33) has already been discussed in Section 10.1. We now see that the coupling strengths are also very similar. Very little is known at the present time about the μ capture process, aside from the absolute decay rate.[161] Certain experiments, such as the directional asymmetry of ejected neutrons following the capture of polarized μ-mesons by nuclei,[162] can in principle yield information about the μ capture interaction equivalent to that obtained by the various "parity" experiments in nuclear beta decay. In general, however, the effects are small and sometimes obscured by specifically nuclear final state interactions, so that details of the μ capture Hamiltonian may be difficult to obtain.

---

[160] J. Tiomno and J. A. Wheeler, *Rev. Mod. Phys. 21*, 153 (1949).
[161] Sens, Swanson, Telegdi, and Yovanovitch, *Phys. Rev. 107*, 1464 (1957).
[162] H. Überall, *Nuovo Cimento, VI*, 533 (1957).

10.4. Conservation of Leptons

The principle of conservation of leptons (or leptonic charge) is completely analogous to the idea of conservation of baryons (the number of heavy fermions minus the number of heavy antifermions is always conserved in any known process). A leptonic charge of $+1$ or -1 is assigned to each of the leptons involved in decay processes. Baryons and bosons are assumed to have leptonic charge 0. In any decay reaction the total leptonic charge or lepton number before and after the decay can be compared to see whether leptonic charge is conserved.

In ordinary beta decay, the question of conservation of leptons is a matter of definition. Since we call the negaton a particle, it is natural to assign it leptonic charge $+1$. We can conserve leptons by calling the "neutrino" accompanying negaton emission an antineutrino with leptonic charge -1. If we believe that the dominant Gamow-Teller interaction is axial vector, we know from the parity experiments that this antineutrino is a right-handed screw. The left-handed neutrino is given the leptonic charge $+1$, while the positon has -1. In μ-meson decay, both ν and $\bar{\nu}$ are emitted. In the reaction $\mu^- \rightarrow e^- + \nu + \bar{\nu}$ the leptonic charge after the decay is $+1$. If we *define* the μ^- to have leptonic charge $+1$ then again we have conservation of leptons by *definition*.

Only in some other process like μ capture or pion decay, for example, can there be physical content to the idea. Consider pion decay within the framework of two-component neutrino theory. If we have conservation of leptons, then a left-handed neutrino is emitted with μ^+. To conserve angular momentum the positive μ-meson's spin is antiparallel to its momentum. In the subsequent μ decay the high-energy positons emerge in the backward direction relative to the μ-meson's line of flight. Adding up the projections of the spins we see that the positons will be longitudinally polarized parallel to their momentum, as was found experimentally ($\xi \simeq -1$). Consequently, if one believes that in ordinary beta decay the dominant Gamow-Teller interaction is axial vector and the subsequent identification of ν as a left-handed screw, one concludes that leptonic charge is conserved in pion decay.[155]

Another source of information about conservation of leptonic charge is double beta decay and inverse reactions (Section 9.1 (e) and (f)). Clearly the process of double beta decay with the emission of two electrons and *no neutrinos* (described as the Majorana case in Section 9.1) violates conservation of leptonic charge, while the unobservable process with four leptons emitted (Dirac case) does conserve leptonic charge. The experimental evidence favored the conservation law. The inverse reactions are in the same category, the conservation law being applied to the entire process of initial emission of the neutrino in ordinary beta decay and its subsequent absorption in the inverse reaction. The negative experiment by Davis, together with the positive result of Cowan and Reines, confirmed the idea of conservation of leptons.

The results on double beta decay and inverse reactions have always been described as "yes or no" propositions. Only recently has Pauli[163] pointed out that, completely aside from parity nonconservation, the Majorana and Dirac cases described in Section 9.1 are merely the two extremes of the situation, and any intermediate circumstance is possible. Thus, even in the parity conserving theory with Dirac neutrinos, conservation of lepton number can be violated to any desired degree with a nuclear beta interaction of the form:

$$H_{\text{int}} = \sum_j (\bar{\psi}_p \mathcal{O}_j \psi_n)[C_{\text{I}j}(\bar{\psi}_e \mathcal{O}_j \psi_\nu) + C_{\text{II}j}(\bar{\psi}_e \mathcal{O}_j \psi_\nu^c)] + \text{h.c.}\,, \qquad (10.21)$$

where ψ_ν and ψ_ν^c are charge conjugates of one another (ψ_ν describes the destruction of a neutrino; ψ_ν^c, the destruction of an antineutrino). The maximum violation will occur for $|C_{\text{I}i}| = |C_{\text{II}i}|$. This observation of Pauli shows that the question of conservation of leptonic charge is not merely a qualitative "either-or" problem, but rather one needing difficult quantitative experimentation.

If we now allow the possibility of parity non-conserving terms, the interaction takes the general form:

$$\begin{aligned} H_{\text{int}} = \sum_j (\bar{\psi}_p \mathcal{O}_j \psi_n)[&C_{\text{I}j}(\bar{\psi}_e \mathcal{O}_j \psi_\nu) + C'_{\text{I}j}(\bar{\psi}_e \mathcal{O}_j \gamma_5 \psi_\nu) \\ &+ C_{\text{II}j}(\bar{\psi}_e \mathcal{O}_j \psi_\nu^c) + C'_{\text{II}j}(\bar{\psi}_e \mathcal{O}_j \gamma_5 \psi_\nu^c)] + \text{h.c.}\,, \end{aligned} \qquad (10.22)$$

replacing (9.10) of Section 9.2. All the results of ordinary beta decay now have added contributions from the terms involving C_{II} and C'_{II}, so that the coefficients of the various vector combinations are generalized in the obvious way by addition of the same combinations of coupling constants with subscripts II as occur with subscripts I.

The question of conservation of leptonic charge is more complex when there are even and odd terms in the interaction. For example, the two-component theory of the neutrino can be viewed as a description of a Majorana neutrino, as mentioned in Section 9.4. Then it is easy to show that conservation of leptonic charge can only occur if the even and odd parts of H_{int} are equal in magnitude, i.e. that ψ_ν occurs in all interaction terms in the combination $(1 + \gamma_5)\psi_\nu$, or as $(1 - \gamma_5)\psi_\nu$, throughout. In physical terms this means that the interaction is such that only one spin state of the Majorana neutrino is operative in the emission or absorption process. If only left-handed neutrinos are emitted, then only right-handed neutrinos can be absorbed to produce the same reaction (not the inverse) and the virtual neutrino in double beta decay cannot be reabsorbed.

---

[163] W. Pauli, *Nuovo Cimento, VI*, 204 (1957).

The decay of K-mesons affords another test of the conservation of leptonic charge. In the $K_{\mu 2}$ mode of decay ($K^+ \rightarrow \mu^+ + \nu$) the directional asymmetry of the electrons from the subsequent μ-meson decay is in the same sense (more electrons backwards relative to the line of flight of the μ^+) as in pion decay.[164] Consequently, we can conclude that leptonic charge is conserved in the $K_{\mu 2}$ decay.

10.5. Pion Decay

The charged pions[165] decay into a μ-meson and a neutrino with a lifetime of 2.6×10^{-8} sec (see Table 1). The question arises as to whether this decay process can be connected to the universal Fermi interaction in any way. The simplest connection, aside from a direct $\pi(\mu\nu)$ coupling

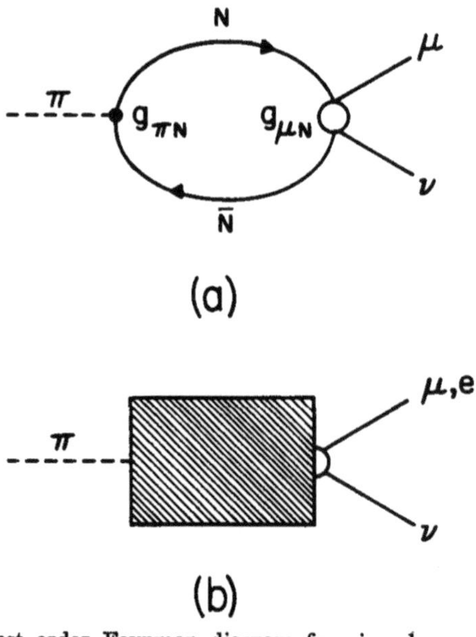

Fig. 21. (a) Lowest order Feynman diagram for pion decay via virtual nucleon pairs.
(b) Pion decay in terms of a black box describing all the strong interactions and one weak vertex.

with another arbitrary coupling constant, is through the strong pion-nucleon (or pion-baryon) interaction (Section 7.3). The pion decay is then envisioned as proceeding through virtual nucleon (or baryon) pair production, followed by ordinary beta decay, the simplest Feynman

---

[164] Coombes, Cork, Galbraith, Lambertson, and Wenzel, *Phys. Rev.* *108*, 1348 (1957).

[165] The neutral pion has a very short lifetime for decay by electromagnetic interaction into two photons. We will not discuss its decay here.

diagram being shown in Fig. 21a. The decay rate calculated from this diagram involves a divergent integral over intermediate states. If this integral is arbitrarily cut off at about the nucleon mass, the decay rate for the pion is of the observed magnitude, provided the $g_{\mu N}$ coupling constant (10.15) is used along with the usual strong coupling constant ($g_{\pi N}^2 \sim 15$). Such a calculation cannot be trusted to give more than a qualitative indication of the behavior expected since the pion-baryon coupling is very strong.

Fortunately something can be said about pion decay and the Fermi interaction, independent of the complications due to the strong pion-baryon coupling. The only assumption is that the weak beta decay interaction is treated only in lowest order of approximation. Fig. 21b shows the phenomenological diagram with a pion entering a "black box" where anything may happen, and a pair of leptons emerging at the other end. The details of what goes on in the black box do not concern us. The pseudoscalar property of the pion restricts the effective invariants in the Fermi interaction to axial vector and pseudoscalar. A choice between these two possibilities, independent of the evidence from μ decay and nuclear beta decay, can be made on the experimental absence of a decay mode $\pi \to e + \nu$. On the basis of the Tiomno-Wheeler triangle we would a priori expect the $\pi \to e + \nu$ mode to compete very favorably with the $\pi \to \mu + \nu$ mode. If ρ is defined as the ratio of the rates for these two modes:

$$\rho = \frac{\pi \to e + \nu}{\pi \to \mu + \nu}, \tag{10.23}$$

the pseudoscalar interaction gives a ρ value:

$$\rho_P = \left(\frac{m_\pi^2 - m_e^2}{m_\pi^2 - m_\mu^2}\right)^2 \simeq 5.4, \tag{10.24}$$

whereas the axial vector interaction[166] has:

$$\rho_A = \left(\frac{m_e}{m_\mu}\right)^2 \rho_P \simeq 1.3 \times 10^{-4}. \tag{10.25}$$

The pseudoscalar value is completely excluded by the lack of any evidence for the electron mode of decay. Unfortunately, the most recent[167] experimental limit on ρ ($\rho < 10^{-5}$) seems also to exclude the axial vector value as well. Disregarding this last point, we find that qualitatively the pion decay is consistent with the Tiomno-Wheeler triangle. We infer, in fact,

[166] The pseudoscalar value for ρ is just the ratio of the densities in phase space. The axial vector value reflects a property of the interaction that inhibits the emission of two relativistic particles back to back. In this case the electron is very relativistic, while the μ-meson is not.

[167] H. L. Anderson and C. M. G. Lattes, *Nuovo Cimento*, *VI*, 1356 (1957).

that the μ capture interaction must be of the same form as the μ decay and nuclear beta interactions, or at least have axial vector coupling present.

Another possible mode of decay of the pion is the radiative process:

$$\pi \rightarrow e + \nu + \gamma . \tag{10.26}$$

An experimental limit of less than 10^{-5} radiative electron decays per ordinary decay has recently been established.[168] This is consistent with the rate of less than 10^{-7} per ordinary decay for axial vector coupling.[169] It should be mentioned, however, that if the nuclear beta interaction involved considerably larger amounts of tensor coupling than axial vector coupling, the theoretical *radiative* decay via tensor coupling[169] relative to the ordinary decay via axial vector coupling would be much larger than is consistent with the experimental upper limit. This argument can perhaps be inverted to lend support to the nuclear beta interaction (9.33) and the idea of a universal Fermi interaction with (V–A) as the dominant invariants, provided we accept the mechanism of pion decay via nucleon pairs. The sole difficulty is the discrepancy between the calculated ρ value (10.25) and the experimental limit of less than 10^{-5}.

10.6. Parity Nonconservation in Hyperon Decay

Since it was the peculiarities of the decay of K-mesons that prompted the investigation into parity nonconservation, it is of interest to see what direct evidence exists from the K-meson and hyperon decay processes. It has already been mentioned that parity is not conserved in the $K_{\mu 2}$ mode of decay of the charged K-mesons. For Λ^0 hyperons there is now evidence for violation of space inversion symmetry in the decay

$$\Lambda^0 \rightarrow p + \pi^- . \tag{10.27}$$

The observed effect is a directional asymmetry of the type found for Co^{60} decay. The Λ^0 hyperons are produced by the interaction of negative pions with protons:

$$\pi^- + p \rightarrow \Lambda^0 + K^0 . \tag{10.28}$$

If outgoing partial waves of orbital angular momentum greater than zero enter in the production, there is a possibility of appreciable polarization of the Λ^0 perpendicular to the production plane defined by $(\vec{k}_\pi \times \vec{k}_\Lambda)$. If parity is not conserved in the decay process (10.27), there can be a directional asymmetry in the decay pions relative to the axis of polarization of the Λ^0. This corresponds to an "up-down" asymmetry relative to the plane of production. In the center of mass of the Λ^0 the directional asymmetry is given by:

$$\frac{dw}{d\Omega} \sim 1 + \alpha \langle \vec{\sigma}_\Lambda \rangle \cdot \frac{\vec{p}_\pi}{p_\pi} , \tag{10.29}$$

[168] Cassels, Rigby, Wetherell, and Wormald, *Proc. Phys. Soc. 70*, 729 (1957).
[169] S. B. Treiman and H. W. Wyld, Jr., *Phys. Rev. 101*, 1552 (1956).

where $\langle\vec{\sigma}_\Lambda\rangle$ is the average polarization of the Λ^0 (perpendicular to the production plane), $\vec{p}_\pi$ is the momentum vector of the decay pion, and α is the asymmetry parameter.[170]

Experimental data are available showing a large directional asymmetry for Λ^0 decays following production by approximately 1 Bev pions according to (10.28). The averaged product of polarization times asymmetry is observed[171] to be:

$$\alpha\left|\langle\vec{\sigma}_\Lambda\rangle\right| = 0.44 \pm 0.11\ , \tag{10.30}$$

showing a large asymmetry. Since the polarization in production is very likely not complete, this value of $\alpha\left|\langle\vec{\sigma}_\Lambda\rangle\right|$ indicates a value of α appreciably greater than 0.5.

Similar effects may be expected for other hyperon production and decays. However, with Σ^- hyperons produced by the same energy pions in the reaction:

$$\pi^- + p \rightarrow \Sigma^- + K^+, \tag{10.31}$$

very little directional asymmetry is found. The reason is not clear, but it may be that at that pion energy the Σ^- are much less polarized in production than the Λ^0. In any event, parity nonconservation is found in Λ^0 decay. A detailed understanding of the magnitude of the effect must await information on the production process, as well as the decay mechanism.

10.7. *K*-Meson and Hyperon Decay Mechanism and a Universal Fermi Interaction for All Decays

The decay of the *K*-mesons can be envisioned in the same terms as pion decay described in Section 10.5. The only difference is that in order to conserve strangeness the baryon pair into which the *K*-meson transforms must involve at least one hyperon. In the subsequent weak decay process strangeness is not conserved. To explain such decay modes as $K_{\mu 2}$ it is necessary to postulate that the Tiomno-Wheeler triangle (Fig. 20) be generalized to allow Fermi couplings of the type $(\bar{\Lambda}N)(\bar{\nu}\mu)$, etc.

Feynman and Gell-Mann,[172] and independently Sudershan and Marshak[173] and Sakurai,[174] have proposed a universal Fermi interaction for all decay processes. They postulate that in the quadrilinear Fermi form *all* fermion fields should be multiplied by $\frac{1}{2}(1 + \gamma_5)$, and that there be one

[170] With spin $\frac{1}{2}$ for the Λ^0, parity nonconservation allows *s*-wave and *p*-wave emission of the pion (Section 6.5). If *S* and *P* are the relative amplitudes for *s*-wave and *p*-wave emission, $\alpha = 2\,\mathrm{Re}\,(SP^*)/(|S|^2 + |P|^2)$.

[171] Crawford, et al, *Phys. Rev. 108,* 1102 (1957), and Eisler, et al, *Phys. Rev. 108,* 1353 (1957).

[172] R. P. Feynman and M. Gell-Mann, *Phys. Rev. 109,* 193 (1958).

[173] E. C. G. Sudarshan and R. E. Marshak, *Phys. Rev. 109,* 1860 (1958).

[174] J. J. Sakurai, *Nuovo Cimento, VII,* 649 (1958).

universal coupling constant G appropriate to vector coupling.[175] A general Fermi interaction would then be:

$$H_{\text{int}} = G(\bar{\varphi}_1\gamma_\mu\varphi_2)(\bar{\varphi}_3\gamma_\mu\varphi_4)\,, \tag{10.32}$$

where

$$\varphi_i = \tfrac{1}{2}(1 + \gamma_5)\psi_i. \tag{10.33}$$

For nuclear beta decay the interaction (10.32) corresponds to the Hamiltonian (9.33) for (V − A), with $G = 4C_V = -4C_A$. In terms of the O^{14} coupling constant (9.4), the Feynman-Gell-Mann coupling strength is $G = \sqrt{8}g_F$.

In μ-meson decay (10.32) corresponds to the interaction (10.1), with $G = 2f_V = -2f_A$. The magnitudes of the μ decay and beta decay coupling constants, $g_{\mu e}$ and g_β are now exactly related by $g_{\mu e} = \sqrt{2}g_\beta$. With the value of g_β from (9.4) and the μ decay result of (10.20), this relation holds to a surprising accuracy and gives striking support for an exact universality among the Fermi interactions. In nuclear beta decay, μ decay, and pion decay the present universal interactions, just as the forms deduced more empirically in Sections 9.10 and 10.1, agree with all the various experimental data, except for the ratio of electronic to μ-mesonic pion decays.

To discuss the K-meson and hyperon decays, Feynman and Gell-Mann postulate that in addition to the usual brackets:

$$(\bar{p}n), \quad (\bar{\nu}e), \quad (\bar{\nu}\mu), \tag{10.34}$$

which are taken in pairs to form the various beta decay, μ decay and μ capture interactions according to (10.32), there are other brackets, such as:

$$(\bar{p}\Lambda^0), \quad (\bar{p}\Sigma^0), \quad (\overline{\Sigma^-}n), \quad (\overline{\Xi^-}\Sigma^0)\,, \cdots \tag{10.35}$$

formed according to the rule, $\Delta Q = \pm 1$, $\Delta S = \pm 1$. This rule is arranged to forbid decay processes with $|\Delta S| > 1$, as seems to be required by all present data (Section 6.1). Use of the general form (10.32) and (10.33) assures parity nonconservation in all decay processes.

With the decay interactions given by various products of the brackets in (10.34) and (10.35), the hyperon and K-meson decay processes can be discussed at least qualitatively along the lines of the pion decay in Section 10.5. For example, the Λ^0 decay (10.27) can be imagined to involve a weak Fermi interaction:

$$H_{\text{int}} = G(\bar{\varphi}_p\gamma_\mu\varphi_\Lambda)(\bar{\varphi}_n\gamma_\mu\varphi_p), \tag{10.36}$$

in which a Λ^0 transforms into a proton and a nucleon-antinucleon pair. The pair then annihilate into a negative pion by strong interaction. The simplest Feynman diagram for this process is shown in Fig. 22.

For K-meson decay involving the emission of leptons, such as $K_{\mu 2}$, the

[175] It is rather amusing that after twenty three years the form of the beta interaction should return to the four-vector invariant originally written down by E. Fermi, *Z. für Physik 88*, 161 (1934). The presence of the $\frac{1}{2}(1 + \gamma_5)$ for each fermion is not, of course, in Fermi's original paper!

lowest order Feynman diagram is similar to Fig. 21 for pions and is shown in Fig. 22. The absence of a K_{e2} mode of decay can be justified in the same manner as for the pion. In the K-meson decays into pions (the θ mode with two pions, and the τ mode with three), the process involves two primary strong interactions and one weak one. A typical diagram[176] for two pion decay is shown in Fig. 22.

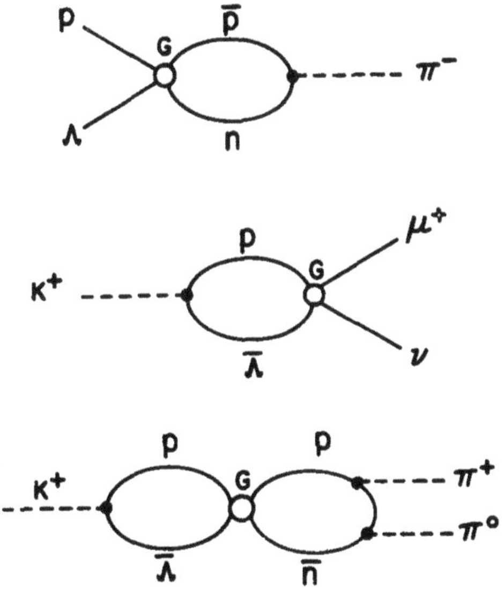

Fig. 22. The upper figure represents the decay $\Lambda^0 \rightarrow p + \pi^-$. The open circle vertex marked G is the weak Fermi interaction at which parity nonconservation occurs. The strong interaction is indicated by a black dot. The middle figure shows a simple diagram for $K^+ \rightarrow \mu^+ + \nu$. The bottom diagram is a typical one for the two pion decay of a K-meson.

By means of the generalized universal Fermi interaction of Feynman and Gell-Mann all decay processes can be understood at least qualitatively. For the well known processes in the Tiomno-Wheeler triangle the theory is in good quantitative agreement with experiment (with the exceptions already noted). Whether the universal quadrilinear Fermi interaction (10.32) is really fundamental, or only an approximation to a second order process involving the exchange of a heavy charged vector meson, is one of the many questions still to be answered. Perhaps the indications of non-local effects in μ decay (Section 10.1) show that the Fermi interaction is an approximation valid only at relatively low energies. If this is true, study of the more energetic K-meson decays should bring to light more definite departures from the simple local theory.

---

[176] There are, of course, many other diagrams, even in lowest order, since the K-mesons are coupled to all the baryons, and pions can be emitted before the weak interaction as well as after.

INDEX

Lightning Source UK Ltd.
Milton Keynes UK
UKHW021955220422
401921UK00006B/370

9 780691 626581